CLIMATE IN CRISIS

THE GREENHOUSE EFFECT
AND WHAT WE CAN DO

By

ALBERT K. BATES

AND THE SPECIAL PROJECT STAFF OF

PLENTY-USA

THEY HAVE SOWN THE WIND,
AND THEY SHALL REAP THE WHIRLWIND.

—*Hosea 8:7*

THE BOOK PUBLISHING COMPANY
SUMMERTOWN, TENNESSEE

To Albert W. Bates
February 3, 1908 - July 15, 1989

CLIMATE IN CRISIS. Copyright © 1990 by The Book Publishing Company. All rights reserved. Printed in the United States of America. No part of this book may be used or reproduced in any manner whatsoever without written permission except in the case of brief quotations embodied in critical articles and reviews. For information address The Book Publishing Company, P.O. Box 99, Summertown TN 38483.

FIRST U.S. EDITION

Library of Congress Cataloging-in-Publication Data
Bates, Albert K., 1947-
    Climate in Crisis: The Greenhouse Effect and What We Can Do / by Albert K. Bates and the special project staff of Plenty USA.—1st U.S. ed.
    p.   cm.
    Includes bibliographical references.
    ISBN 0-913990-67-1: $11.95
    1. Global Warming. I. Plenty-USA. II. Title.

QC981.8.G56B38 1989
Dewey Dec: 363.73'92—dc20
Library of Congress Catalog Card Number   89-17890
    CIP

Climate in Crisis

Project Staff

Project Director: Albert K. Bates • Publishers: Robert and Cynthia Holzapfel • Contributing Editors: Albert W. Bates, Dorothy R. Bates, Robert Bransome, B. Dolph Honicker, Richard A. Houghton, Kathryn Hill, Bill McNew, Frank Michael, Carol Nelson, Peter Schweitzer, Dan and Rachel Sythe • Research Staff: Josh Amundson, Gretchen Bates, Eva Gaskin, Ina May Gaskin, Pagan Hill, Mary Hubbard, Gabe Hurgeton, Michael Hurgeton, Kachina McMurry, Mary Moore, Esther Traugot, Vivian Traugot, and Deborah Weinishke • Design, Text, and Illustrations: Albert K. Bates • Typesetting by Jim Bailey at Bailey Typography • Cover by Barbara McNew and Eleanor Dale Evans.

Acknowledgements

The author wishes to acknowledge the kind assistance of the many people who generously contributed their time and energy to assist as researchers and editors. In addition, special gratitude is due to Cynthia Bates, Peter Berg and Judy Goldhaft, Jaz Bertolette, Cooper and Leimomi Brown, Lester Brown, James McClure Clark, Douglas Cobb, Claudine Cremer, Karen Flaherty, Eve Furchgot, Stephen and Ina May Gaskin, Bart Gordon, Albert Gore, Jr., Way Konigsberg, Dorothy Legarreta, John Lemons, Amory and Hunter Lovins, Joshua Mailman, Arjun Makhijani, Bill McKibben, Jessica Mitford, William Prescott, Jeremy Rifkin, Stephen Schneider, June Sythe, Vince and Sunshine Taylor, Dennis Tirpak, Thomas and Lisa Wartinger, and many others, from underpaid environmental activists and underacknowledged government employees to understaffed academics who provided comments, materials, resources, and inspiration. Thanks, too, to Aldus Corporation for providing Freehand, to Jasmine Technologies for supplying a Direct Drive 90, and to Abacus Planned Community Project for wide-ranging assistance. This book was written and compiled on Apple Macintoshes with the aid of Aldus Freehand, Aldus PageMaker, Microsoft Word, and Microsoft Works and output by Linotronic 300. Type is Avant Garde on Goudy with Goudy Outline and Zapf Chancery in the chapter title illuminations. The project staff is grateful to the assistance of all who saw the value of the work and lent a helping hand.

CLIMATE IN CRISIS
TABLE OF CONTENTS

Foreword by Senator Albert Gore, Jr./vii

PART ONE

WAKE UP CALL/3

THE GREENHOUSE CENTURY/5
A Greenhouse Primer/7, Goldilocks and the Three Planets/8,
Trace Gas Chemistry/11, The Ozone Hole/16, Acceleration/17,
Deforestation/18, Changing Patterns/19,
Dealing with Uncertainty/21

RUNAWAY/25
Genesis of Earth's Atmosphere/26,
Axial Tilt, Solar Cycles, and Ice Ages/28,
The Carbon Cycle/29, The Cycles of the Sun/32,
The Runaway Reaction/33, Chaos/34, What if .../38

THE RISING TIDE/41
Ocean Seasoning/44, Sea Level Rise/45,
Changing Shorelines/47, Dikes and Sea-Walls/48,
Channelization/52, Beaches/53, Estuaries/55, Wetlands/56,
Rise of Sea Level in the Third World/57,
An Ocean Never Forgets/60

SUMMER HEAT/61
The Great Plains/64, Monsoons and Typhoons/70,
Summer in the City/71, Increased Mortality/72,
Planning for Change/74

THE SKY IS FALLING/75
Probing the Causes/77, Chlorine/78, Ozone/80,
Health Effects of Ultraviolet Radiation/82, Phasing Out/84

TUMBLING DOWN/87
Deforestation/90, Extinction/96, The Race of the Rooted Ones/97

HUMAN DIMENSION/103
Evolution/104, The Need for Food/108, Over the Crest/113,
Real Food for Real People/114, The Crisis Point/115

PART TWO

A SHIFT IN EMPHASIS/125
Restructuring/127, Getting it Together/128

DEEP ECOLOGY/131
Redefining Natural Rights/137

THE NEW AGENDA/139

TWENTY-ONE BETTER IDEAS/193
Population/148, Energy/156, Food Security/174, Armaments/179,
Reforestation/183, Individual Choices/185

THE NEW AGENDA/189

PART THREE

Afterword by Peter Schweitzer/193

SELECTED REFERENCES/196
GLOSSARY OF TERMS/218
INDEX/224

ABOUT PLENTY ...

PLENTY USA is a non-profit, volunteer, international, relief, development, environment, education, and human rights organization. Albert Bates is director of PLENTY's environmental law project, The Natural Rights Center, in Summertown, Tennessee. For more information about our work, contact PLENTY USA, Davis, California 95617-2306 (Tel: 916-753-0731; MCI-Mail: PlentyUSA).

An explanation of why and how PLENTY, and this book, came into being is contained in the *Afterword* by Peter Schweitzer.

PLENTY's efforts in sustainable development, environmental protection, and public education are supported entirely by private donations.

Foreword

Albert Gore, Jr.
Member, United States Senate

When I was born, in 1948, the population of the world was around 2 billion. It took 10,000 human lifetimes for the population to reach 2 billion. Now, in the course of one lifetime—yours and mine—it is rocketing from 2 billion to 10 billion, and is already halfway there.

If you were to graph the growth of human population, you would see that the graph runs along fairly evenly until recent times and then it starts bending upwards. The same kind of graph would fit well for levels of carbon dioxide in the atmosphere, methane production, and emissions of other greenhouse gases to our atmosphere. Today the graph runs steeply upwards with no end in sight. We get substantially similar results by graphing our loss of the ozone shield, loss of species diversity, loss of forests, erosion of topsoils, pollution of the oceans, and generation of garbage. And yet, so far, the pattern of our political thinking remains remarkably unchanged.

In 1988, I went to the coldest place on Earth to study global warming. High in the mountains of Antarctica, a badly sunburned scientist talked to me about the ice cores he and his team were pulling from a deep hole drilled into the glacier on which we stood. These and other ice cores contain highly accurate information concerning the makeup of the Earth's atmosphere year by year for tens of thousands of years. From such cores, we know, for example, that there were dramatically lower levels of carbon dioxide in the atmosphere at the peak of the last Ice Age. By contrast, ice and snow laid down in the 1980s show levels of carbon dioxide, methane, CFCs, nitrous oxide, and other gases responsible for the greenhouse effect higher than they have been in at least 160,000 years, as far back as the ice cores measure.

In 1989, I visited one of the warmest places on Earth, the Amazon jungle, where an area of rainforest the size of a football field is cut down or burned every second. At current rates of deforestation, it is only a matter of time before most if not all of the world's tropical rainforests, and all of the myriad species of living things that inhabit them, are destroyed. If we let it, this catastrophe could well happen within our lifetimes.

The fact that we face an ecological crisis without precedent in historic times is no longer a matter of any dispute. Those who, for the purpose of maintaining balance in debate, take the contrary view that there is significant uncertainty about whether it is real are hurting our ability to respond.

Seizing upon uncertainties as excuses for inaction is a psychological problem which seems to be common, not only among humans but in other animals as well. There is an old science experiment in which a frog is put into a pan of water, and the water is slowly heated to boiling point. The frog sits there and boils because its nervous system cannot react to the gradual increase in temperature. Of course, if you boil the water first and then put the frog in, it immediately jumps out.

Today we humans are reaching an environmental boiling point. What will it take for us to react? If, as in a science-fiction movie, we had a giant invader from space clomping across the rainforests of the world with football field-size feet, would we react? That's essentially what is going on right now, and yet we are hardly moving. We seem to possess an inherent unwillingness to believe that something so far outside the bounds of historical experience can, in fact, be occurring.

Today in both the national and international arena, short-term policies predominate over long-term policies. Politics is characterized by actions to confer national advantage at the expense of actions designed to promote global advantage. Bit by bit we are coming to realize that short-sighted, narrow-minded thinking will not suffice for preserving the natural resource base upon which our society depends. Yet we cannot expect some sudden transformation of human nature, by which people will be endowed with new virtues of foresight and restraint. We have to rely on the virtues we already have. Jefferson's educated citizenry will have to suffice. The changes demanded of us will be motivated by our natural instinct to survive, and by our often even more intense desire to provide for the survival and well-being of our children and grandchildren.

Only now are we beginning to see, in Ivan Ilych's phrase, "the shadows our future throws." We can influence the shape those shadows take by what we decide to do, but first we must know where to begin. Perhaps the largest difficulty we face in reacting to the current environmental crisis is the lack of widespread awareness among the peoples and leaders of the world about the nature and extent of the problem. For this reason, *Climate in Crisis* is a valuable addition to the public debate.

As individuals and leaders we must shift the world's political system into a new state of equilibrium—an *eco-librium*—characterized by greater cooperation and by a focus on the future. The solutions we seek will be found in a new faith in the future of life on Earth after our own, a faith in the future that justifies sacrifices in the present, a new moral courage to choose higher values in the conduct of human affairs, and a new reverence for absolute principles that can serve as stars by which to map the future course of our species and our place within creation.

Part One

Chapter One
WAKE UP CALL

WAKE UP!

You are falling, falling, falling. The ground is rushing up at you. It's too late to do anything except ... WAKE UP! It was just a dream.

Or was it?

Humans are a curious sort. They seem to relate to their surroundings in a thoughtful way. They communicate exceedingly well. They can cooperate for their mutual benefit. They can adapt, improvise, invent.

An elastic vocal cord, a curved spine, an opposable thumb and a few other minor improvements have given the human animal power over all other things on Earth. In the space of a few millennia, these chattering, hairless monkeys managed to construct something called civilization.

Amazing thing, this civilization. Bristling towers of fused sand and stone. Highways of artificial rock through mountains and across swamps. Aircraft that traverse an entire continent or ocean in the space of a symphony.

Today we don't just signal to each other with smoke or drums. We have satellites in orbit transmitting voices, visual images, and computer data among people of different cultures and languages; satellites taking readings of the Earth's ocean temperatures, ice flows, radiation levels, and storm patterns; space platforms looking at the sun, the planets, and the most distant stars we can see.

Some of us are sitting at microscopes staring at the DNA molecule, labeling genes that cause cancer, or comparing genetic codes to determine how we evolved. Some of us are looking out to the most distant regions of the universe, toward the light from the Big Bang, while others are examining the stuff that elemental particles are made of. We can measure things that are so far distant as to strain the relationship of time and space or are so tiny that we know of them only by their deeds.

We are at the ignition point of an information explosion. The library doors to the physical universe are swinging wide, enabling us to read about everything from our own molecular genesis to the music of the spheres. And it has all been made possible by our ability to communicate.

So listen: We are falling, falling, falling.

The point of this book is to put a strong message out into that enormous, fast-paced global conversation. The ground is rushing up at us. We have to WAKE UP! This is not a dream.

Chapter Two
THE GREENHOUSE CENTURY

y January of 1989, the scientific proof was in: Nineteen Eighty-eight had surpassed the previous record for the hottest year in history by 0.02°F. For the previous record-holder, we had to look all the way back to Nineteen Eighty-seven.

The top six warmest years in the history of recorded temperatures were 1988, 1987, 1983, 1981, 1980 and 1986, in that order. Six records in nine years.

In fact, the 20th century was one degree warmer than the 19th century.

One degree might not seem like much, but that is a worldwide average. The one degree increase worldwide meant very little change in some areas, such as near the equator, but it brought about substantially warmer summers—several degrees on average—at the most northern and southern latitudes.

**Average temperature worldwide
over the past 120 years
in degrees Celsius**

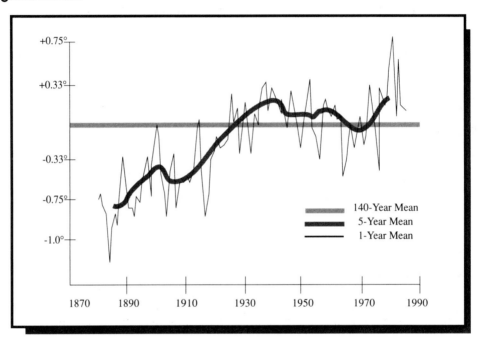

Source: *University of East Anglia, 1987.*

To get a one degree increase, you have to heat a lot of ocean water and a lot of atmosphere. It took 18,000 years for the Earth to warm 9 degrees Fahrenheit (5° Celsius) from the last ice age to the present, about one-half degree every thousand years. The fastest the Earth has previously warmed was only about 3.6°F (2°C) per thousand years. One degree in a single century is a substantially faster rate of warming, 20 times faster than the average.

Yet, from what we experienced in the 1980s, and what we know today about atmospheric and oceanic processes, the warming trend appears to be accelerating. The one degree warming we experienced in the 20th century may be just the curtain rising on a natural event of enormous magnitude. At the end of the 21st century the Earth may be warmer than it is now by as much as another 9 degrees (5°C). That would be warmer than it has been in 1,000,000 years.

A little warming is harmless enough; the Earth's air and water have been gradually warming and cooling, warming and cooling, and warming again ever since they came into existence. The Earth has been hotter than it is now, and it will soon be hotter again.

But there are two things that distinguish the current global warming trend: (1) it now seems evident that this most recent warming is caused by human activity, rather than by geological or astronomical processes that were going on before; and (2) the current warming is happening much faster than it has ever happened, looking back millions of years.

Both differences are ominous.

A Greenhouse Primer

All objects with heat give off radiation with different energies and at different wavelengths. The white hot sun gives off high-energy short-wavelength radiation. The much cooler Earth surface gives off longer-wavelength, lower-energy radiation. Earth's atmosphere, like the atmospheres of Mars and Venus, works like greenhouse glass. It allows the short-wave, direct radiation to pass through but traps the long-wave, reflected heat radiation. That's what is known as "the greenhouse effect."

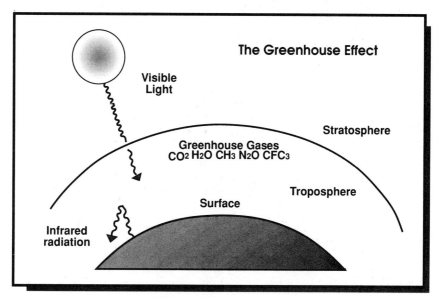

To understand how this works, think of an automobile with all the windows rolled up on a sunny day. The glass windshield admits direct solar energy but traps the radiant energy reflected from the seats and dashboard. The car gets very hot (and very stuffy). Earth's atmosphere works the same way: carbon dioxide, nitrous oxide, methane and other gases tend to trap radiant heat at the Earth's surface—the denser these "greenhouse" gases, the more heat trapped.

About a third of all light reaching the Earth from the sun is reflected back by clouds, ice particles and air molecules. Of the penetrating sunlight, about 25 percent is temporarily absorbed by the atmosphere as warmth but later returned to space, 5 percent is reflected back by the land surface, particularly by ice and snow, and the rest, about 37 percent, is absorbed by the Earth's surface. Most of this absorbed solar radiation is re-radiated skyward during the dark hours of the night, if not sooner, and that is when the greenhouse gases do their work. The greenhouse shield captures nearly 90 percent of the outgoing heat radiation and retains it long enough to keep the planet's temperature in comfortable equilibrium.

GOLDILOCKS AND THE THREE PLANETS

Mars has a thin atmosphere and a surface temperature that is colder than the freezer compartment in your refrigerator (about 65° below zero °F). Venus has a thick atmosphere and is hotter than an oven (about 800°F). Earth has a moderate atmosphere and so contains liquid water and abundant life. Climatologists call this the Goldilocks Phenomenon.

Without the shielding greenhouse provided by the envelope of atmosphere around the Earth, our planet would be 50 to 100°F (27 to 56°C) colder than it is. There would be snow in the Sahara desert and ice on the Amazon river.

What makes up the greenhouse shield? Water vapor and carbon dioxide are roughly half of the shield. Other gases include methane, nitrous oxide, ozone, sulfur dioxide, and a wide array of trace gases (chlorocarbons, hydrocarbons, aldehydes, fluorinated, chlorinated and brominated species, carbon monoxide, nitrogen oxides, and compounds of sulfur).

It is doubtful that human activity could affect the content of oxygen or nitrogen in the atmosphere in any significant way in less than several million years. Those gases have been building for billions of years. However, we have been able to tamper with the greenhouse gases because they are such a small fraction of the Earth's total atmosphere. That small fraction is within the scope of human ability to change quickly and dramatically.

Over the past few centuries, we have been drawing down stored carbon from billion-year-old repositories in oil and coal and from previously rotating stores of forest cover, and sending all of it at once to the atmosphere, much faster than volcanic and forest burning processes that had been going on before. The human population is also generating new greenhouse gases at an unprecedented rate. The equilibrium that the Earth has reached through millions of years of adaptation and delicate balancing is being upset.

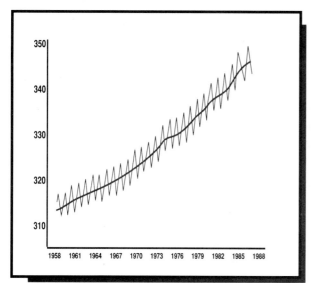

Carbon dioxide in the atmosphere in parts per million measured at Mauna Loa, Hawaii

Carbon dioxide has been rising steadily since a laboratory was established by Charles Keeling in 1958 to take regular measurements at 20-minute intervals. Keeling was hired by the Scripps Institute of Oceanography to determine whether the theories of increased carbon dioxide advanced by Roger Revelle and Hans Suess were correct. The annual oscillation Keeling observed is caused by the changes in absorption of carbon dioxide by seasonal vegetation.

Source: *Scripps Institute of Oceanography.*

Correlation of atmospheric carbon dioxide (CO₂) and methane (CH₄) to global warming and cooling over 160,000 years.

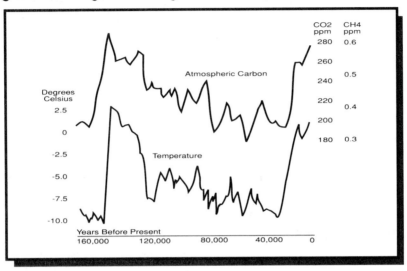

Source: Barnola, et. al., 1987.

By examining the layers of ice laid down each year in Antarctica over the past 160,000 years, scientists are able to conclude with considerable confidence that the carbon content of the atmosphere and the temperature of the Earth are directly related. As far as we can look back in time, as carbon rises, temperature increases. As carbon declines, the Earth's climate cools.

It is only very recently that we have started to realize our plight. It was not until 1938 that British scientist George Callendar, drawing upon earlier speculation by Svante Arrhenius and Jean-Baptiste-Joseph Fourier, offered his hypothesis that substantial greenhouse warming might result from fossil fuel burning; not until 1958 that Charles Keeling began to measure atmospheric carbon at the Mauna Loa observatory, and only in the last decade that the steep upward trend has become too obvious to ignore; not until 1984 that a British expedition led by Joseph Farman discovered a massive hole in the ozone shield over Antarctica; not until 1985 that French and Soviet scientists studying ice cores reported a correlation between atmospheric carbon and temperature that could be traced back to prehistoric times; not until 1987 that American climatologist James Hansen at the Goddard Institute of Space Studies reported that, by combining satellite images, surface, air, and ocean data and using a supercomputer, it could be said with almost complete certainty that the Earth was rapidly heating.

Since the dawn of the industrial era, only about 200 years ago, the upper atmosphere has been changing as a result of human activities. In the past century, carbon dioxide levels have increased by 25 percent due to the burning of fossil fuels (oil, gas and coal) and the destruction of forests. Methane in the upper atmosphere has more than doubled, primarily from the venting of oil and gas wells, but also as a result of conversion of forests and fields to cattle production and rice paddies, the harvesting of the oceans for fish, and the decay of organic wastes from those activities. Atmospheric nitrous oxide, a pollutant emitted by burning coal and oil, denitrification of fertilizer, and deforestation, has increased by one-third.

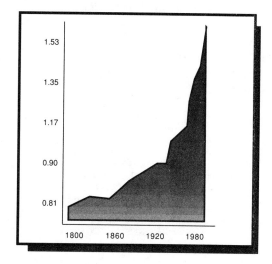

Concentration of atmospheric methane over the past 180 years in parts per million worldwide

Methane is produced by the process of digestion in cattle and domestic animals, the fermentation of rice paddies, and the digestion of forest organic matter by insects and microbes. While methane has been naturally produced since the formation of the Earth, present day rates of production exceed anything in the historical record.

Source: Mintzer, 1988

TRACE GAS CHEMISTRY

Equally important as the major greenhouse gases are the many so-called "trace" gases. While these compounds are present only in fractional quantities—less than one part per million parts of air—some of them are far more powerful in affecting global temperature, atom for atom, than more abundant gases like carbon dioxide (CO_2) and methane (CH_4). Among these more powerful compounds are the chlorofluorocarbons (CFCs) used in refrigerants, aerosol sprays, insulating materials, and solvents. CFCs are exclusively man-made and linger in the upper atmosphere with little degradation for 50 to 200 years. CFC-22—an air conditioning and refrigeration coolant untouched by the recent Montreal accord (a treaty to limit CFC releases that we will look at more closely in Chapter 6)—is doubling its concentration in the upper atmosphere every 8 to 10 years.

The Greenhouse Gases

Greenhouse Gas	Formula	Sources	Years to Degrade
Water Vapor	H_2O	Natural Evaporation	0.001
Carbon Dioxide	CO_2	Fossil fuels and deforestation	8
Nitrogen Compounds	N_2O	Combustion and fertilizer	120
	NH_3	Agricultural chemicals	0.01
	NO, NO_2	Combustion	0.001
Sulfur Compounds	CSO	unknown	unknown
	CS_2	unknown	unknown
	SO_2	Combustion and industrial	0.001
	H_2S	Combustion and industrial	0.001
Fully Fluorinated	CF_4 (F14)	Aluminum industry	>500
	C_2F_6 (F116)	Aluminum industry	>500
	SF_6	unknown	>500
Chlorofluorocarbons	$CClF_3$ (F13)	Air conditioners, refrigerants, aerosols	400
	CCl_2F_2 (F12)		110
	$CHClF_2$ (F22)	" "	20
	CCl_3F (F11)	" "	65
	CF_3CF_2Cl (F115)	" "	380
	$CClF_2CClF_2$ (F114)	" "	180
	CCl_2FCClF_2 (F113)	" "	90
Chlorocarbons	CH_3Cl	Natural ocean	1.5
	CH_2Cl_2	Industrial solvents	0.6
	$CHCl_3$	Manufacture of F22	0.6
	CCl_4	Manufacture of fluorocarbons	25-50
	CH_2ClCH_2Cl	Chemical industry	0.4
	CH_3CCl_3	Degreasing solvent	8
	C_2HCl_3	Degreasing solvent	0.02
	C_2Cl_4	Degreasing solvent	0.5
Brominated/Iodated	CH_3Br	Natural sources	1.7
	$CBrF_3$	Fire extinguishers	110
	CH_2BrCH_2Br	Leaded gasoline additive	0.4
	CH_2I	Natural ocean	0.02
Hydrocarbons	CH_4	Industrial	5-10
	C_2H_6	Automobile exhaust	0.3
	C_2H_2	Industrial	0.3
	C_3H_8	Natural sources	0.03
	CO	Widely varied sources	0.3
	H_2	Widely varied sources	2
Tropospheric Ozone	O_3	Natural sources	0.1-0.3
Aldehydes	$HCHO$	Hydrocarbon oxidation	0.001
	CH_3CHO	Natural sources	0.001

Carbon monoxide and nitrogen oxides, like carbon dioxide, are produced by fossil fuel combustion, oxidation of methane, and the burning of savanna, agricultural land, and forests. Carbofluorines such as CF_4 and C_2F_6 are by-products of aluminum manufacture. Scientists predict they will linger in the atmosphere at least 500 years, but do not know exactly how long. Chlorocarbons such as CCl_4 and CH_3CCl_3 are released in the manufacture and use of degreasing solvents such as oven cleaners. They can linger in the atmosphere up to 50 years. Brominated and iodated compounds from fire extinguishers, gasoline additives, and fumigants can last more than 100 years. All these substances have been rapidly increasing in the Earth's atmosphere in the past century.

Slicing the Atmospheric Pie
Relative sources and emissions
of principal greenhouse gases

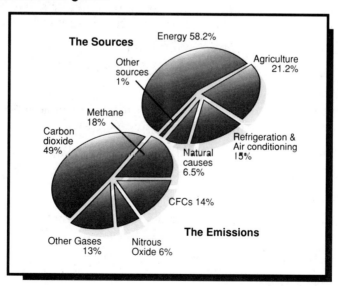

Sources:
Nordhaus, 1989;
Passell, 1989.

A number of the trace gases are greater than 10 times more powerful in absorbing infrared radiation than the more common greenhouse gases. The most powerful are 100 times more potent than methane and more than 1000 times more powerful than carbon dioxide. If we remember our analogy of the heating automobile—in which we heated a large volume of air behind a thin transparent window—we would find we have even greater leverage if we concentrate our efforts on using the most efficient materials in creating that thin enclosure. Because of the size of the Earth's atmosphere, our alterations to air chemistry have been miniscule by volume—they are measured in parts per billion in the upper atmosphere. But by randomly using and discharging a small combination of very effective greenhouse gases, we have created conditions for significant warming.

Probable Effects of Atmospheric Changes

- Global surface warming, with most profound changes occurring farthest from the equator.
- Global precipitation increase, although some regions may become dryer because of rapid evaporation and changing wind patterns.
- Wetter monsoons in the coastal subtropics.
- More frequent and heavier winter snows at high altitude and high latitude.
- Earlier snowmelt, wetter springtime, earlier, longer summers and more frequent droughts in interior, mid-continent regions.
- Improved agricultural conditions in high latitudes.
- Reduction of sea ice, with sea ice boundary shifting northward.
- Coastal sea level rises of several feet per century.
- More frequent and powerful hurricanes, extending to higher latitudes.
- More frequent and severe forest fires.
- Rapid species extinctions.
- Increased human mortality due to weather related causes.

The net effect of man-made releases of greenhouse gases into the atmosphere is to make the greenhouse shield more dense to infrared radiation, so that it traps more heat, and holds it longer. Earth's response is to warm itself to a healthy fever, speeding evaporation and rainfall. A warmer atmosphere retains more moisture. Water vapor absorbs more radiant heat. More infrared absorption makes the atmosphere warmer still. Eventually, all the extra water vapor will make more rain. Hopefully, over the course of many centuries, that rain will wash the excess greenhouse gases out of the atmosphere. That is Earth's natural atmospheric cleansing mechanism. But until the density of greenhouse gases actually diminishes, the planet will get warmer and warmer.

Fevers are a natural process that help a body overcome an infection—if they don't kill the patient. The Earth's fever is just beginning now. It will put great strain on Earth's naturally regenerative systems. For a number of reasons, we can expect things to get considerably worse before they get any better.

Ice and snow reflect visible light back into space. The fraction reflected by the surface is called the Earth's "albedo." Climate warming melts snow and ice, which reduces the albedo and produces even greater warming.

Forests absorb carbon dioxide from the atmosphere in much the same way that humans breathe in oxygen. Global warming, acid rain, and other factors are increasing the rate and extent of forest fires, reducing the amount of carbon dioxide being absorbed while increasing the release of stored carbon to the atmosphere, which leads to still more global warming.

Although more than twice as much carbon is stored in forest wood as is in the atmosphere, about twice that much again is stored in soils. It is stored in the form of dead organic matter, such material as last season's weeds, leaves and insects. As this organic matter decomposes, it gives off its carbon to the atmosphere, usually as molecules of carbon dioxide or methane. Warmer temperatures will increase the rate of decomposition, leading to still more global warming.

Methane in enormous quantities is trapped under frozen tundra in the Arctic. As the tundra thaws, the methane will be released. Methane is a very efficient greenhouse gas—about 30 times more effective per molecule at absorbing infrared radiation than carbon dioxide—so global warming will increase.

THE OZONE HOLE

The importance of aerosol sprays was not recognized until 1974, when chemistry professors Mario Molina and F. Sherwood Rowland suggested that CFCs such as freon could eventually destroy as much as 20 percent of the Earth's ozone layer. Molina and Rowland's dire predictions were greeted by skepticism until British researcher Joseph Farman reported in the journal *Nature* in 1985 that there was a large and growing hole in the ozone over Antarctica. In 1988, an American Antarctic expedition led by atmospheric chemist Susan Solomon confirmed that chlorofluorocarbon propellants were the reason for the hole, just as Molina and Rowland had predicted.

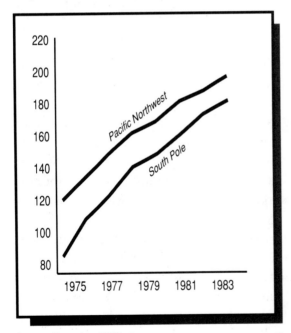

Increase in CFC-11 from 1975-1983 South Pole and Pacific Northwest in parts per trillion by volume

Source: Patrusky, 1989.

Without the ozone shield, ultraviolet rays penetrate to the surface of the oceans. There they destroy the thin-celled phytoplankton that are important to atmospheric chemistry, both as the Earth's greatest absorbers of carbon and as the world's largest producers of oxygen. Because plankton have no layers of skin, they have no protection against ultraviolet radiation.

What will the increase of carbon dioxide and methane and the decrease of ozone mean to us? A doubling of atmospheric carbon by the middle of the next century (from 300 parts per million to 600 parts per million) will likely raise global temperatures between 2° and 9°F (1° to 5°C). If we continue present rates of acceleration, an increase as large as 9°F might be accomplished as early as between 2030 and 2050, 40 to 60 years from now. Considering the fact that the Earth has only warmed 9 degrees in the past 18,000 years, it is difficult to imagine what a 9 degree increase in just 40 years might mean.

Acceleration

Both plants and animals typically react to changing climate by biologically adapting and migrating as temperatures rise and fall. The abruptness of man-made climate change will not permit a gradual biological evolution this time and migration will be difficult to impossible because of the extent of deforestation and human settlement. Two thirds of the world's fertile wetlands will vanish as the oceans rise. Many familiar (and many unknown) species will simply disappear.

There is actually a time delay of at least 30 years from the addition of carbon to the upper atmosphere to the peak change in climate that results. The delay is caused by the time it takes for atmospheric gases to mix and mingle and by the slow response of the oceans and continents to heating. *This delay factor means that the climate we are experiencing now is the result of our industrial activities before 1960.*

Last year, fossil fuel combustion emitted some 5.4 billion tons of carbon into the atmosphere, while deforestation released on the order of 1 to 3 billion tons. In 1960, total carbon emissions were less than one-third of that. Moreover, our industrial and population growth trends are not only much greater now than 30 years ago, they are accelerating. There were only 50 million automobiles in the world in 1960. By the year 2000, there will be 500 million.

Deforestation

The principle way that carbon dioxide is removed from the atmosphere is through absorption by plants. A forest is an atmospheric scrub brush, drawing carbon dioxide from the air, returning oxygen to the atmosphere and locking the carbon away in cellulose fiber. Forests are one of the most efficient means humanity has to restore the atmospheric carbon balance and to delay or reverse greenhouse warming, but it is a tool that is in the process of being discarded.

Deforestation has been going on since the dawn of human agriculture, but it has increased dramatically in industrial times. In the year 900, about 40 percent of the Earth's land mass was forested. In 1900, one thousand years later, only 30 percent was forested. Today, *just one hundred years later*, about 20 percent is forested, and that remaining forest is falling faster than at any time in history.

Most of the great forests in the world today are in the equatorial zones, and because these regions are not periodically glaciated, the soils are typically poor in minerals. The trees are there because for many eons, separated from human conquest by distance and terrain, they have been undisturbed.

By and large, equatorial trees get their nutrients as they arrive from the sky or decay from other trees. If these forests are disturbed, they won't grow back easily. Today the equatorial forests are falling at an astonishing rate; a football field every second, an area the size of France every five years. The full might of multinational human economies is turned toward these forests' destruction. The World Bank, the International Monetary Fund, the Inter-American Development Bank, and other large funding organizations are sponsoring chain-saw crews in order to make room for cattle ranches and other commercial ventures. In just a few years, the rich forest along the Eastern coast of Brazil has been cleared to less than 1 percent of its original cover. The clearing operation underway in Western Brazil is so large its flames are visible from the Space Shuttle in orbit around the Earth.

Deforestation due to acid rain, ozone loss, and forest fires is also increasing, and being accelerated by shifting wind and temperature patterns. These effects are now devastatingly apparent in the forests of Canada, the United States, Scandinavia, West Germany, Eastern Europe, and the Soviet Union.

With the loss of the world's forests goes the accumulated genetic wealth of the ages. Today we are losing from 4,000 to 6,000 species of plants and animals each year, a rate of loss 10,000 times greater than the rate that prevailed before the human era. Since only a few of our fellow species have been studied and most are completely unknown to us, the true cost of their loss to Earth's natural systems and to future generations of humans is incalculable.

CHANGING PATTERNS

Global warming trends will not only increase total rainfall, but change rainfall patterns. Some regions will become wetter, while others will become much drier. Much depends on the reaction of air currents to the changing ocean temperatures and on the changes in ocean currents in response to narrowing temperature differences between the equator and the poles. While meteorological models of local regions are still sketchy, historical evidence suggests that the mid-continent grain-growing regions of North America and the Ukraine are likely to lose moisture. If so, the Great Plains would revert to grassland or become desert, and the North American corn belt would suffer droughts with increasing frequency.

There were great lakes in the American Southwest and vast spruce forests east of the Mississippi at a time when the Northwest was dry. Southward displacement of the jet stream 18,000 years ago brought lakes that filled what are now desert basins into the Southwest. As the jet stream moved northward 12,000 years ago, those lakes dried up. But at the same time that the American Southwest became a vast desert basin, on the other side of the world, the lakes of the Sahara were expanding.

**Soil moisture in North America
with a 4°C increase in temperature**

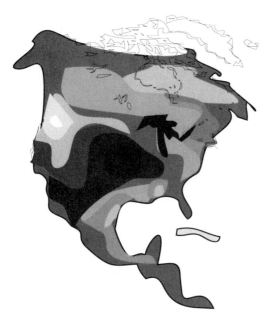

Although precipitation is expected to increase, the increase in temperature will also speed evaporation. As a result, the mid-continental regions which receive less moisture from the oceans will become much more arid, while some coastal areas may become wetter. In this computer-generated model, light areas represent wetter or constant soil moisture and darker areas represent drier soils. In the center of the continent (darkest circle) normal soil moisture is reduced to only half of the present day average.

Source: Hansen, 1988.

In 1500 B.C., droughts in Central Asia forced Aryans from Iran into northwest India and drove horse-riding barbarians to conquer the advanced Yang-shao and Lung-shan cultures in China. In 1200 B.C., the Mycenean empire in southern Greece was overrun and destroyed by Dorian tribes escaping drought-caused famines to the North. The same general warming trend may have driven the Hittites and Syrians into Egypt.

Until 1280 A.D. the Anasazi Indians of Colorado had a flourishing civilization that carved magnificent cities into massive cliff walls and connected remote villages and farms by hundred-mile-long stone roads as straight as an arrow. In 1280 the Anasazi civilization collapsed and vanished, the victim of a decade-long drought. While the Anasazi had weathered similar droughts in prior centuries, in the drought of 1280 they were doomed by their own economic growth. Earlier Anasazi villages had survived by virtue of their limited size and their ability to scratch a subsistence from desert conditions even in meager years. By the 13th century, the Anasazi population had grown to match the carrying capacity of the Mesa Verde region in good years. It could not be sustained through successive bad years.

While people are more able to adapt to changing weather today than they were in earlier centuries, the changes now in store will test the agricultural and water management skills of many nations. As the heat difference between the poles and the equator narrows, the North American jet stream may again migrate southward, refilling the Great Salt Lake with fresh water and giving Arizona the wet climate of the Amazon. It is equally possible that the jet stream could migrate northward, which would reduce the already meager rainfall in arid parts of North America until that region once more resembles the shifting sands of the Sahara. These changes may come to pass before the children born today reach the age of their parents.

DEALING WITH UNCERTAINTY

On a cold day in November, 1987, United States Senator Tim Wirth convened a hearing on the greenhouse effect. Present to testify was James Hansen, a climate modeler with NASA's Goddard Institute of Space Studies. What Hansen had to say, in the quiet demeanor of a well-respected scientist, was that the Earth was getting warmer, there was a high degree of probability that the warming was associated with human-induced augmentation of the greenhouse effect, and that the effects of the change would likely become very pronounced in the 1990s and thereafter.

Hansen's statements created a hush in the hearing room. It was a break in the ranks of the government science community, whose first law is never to draw a conclusion, but if you must, draw it tentatively, sprinkled with abundant qualifiers and caveats about the need for more study. Hansen said he could state his three points with "99 percent certainty."

The trouble was, the only people there to listen were a few Senators, some congressional aides, and one or two tourists who wandered into the committee room by chance. There were no television lights, no print reporters madly scrawling on notepads, no cameras and tape recorders. Senator Wirth, sensing the need to get this startling news to more people, asked what time of the year was likely to be the hottest for Washington, D.C.. Hansen, who was not a weatherman, suggested the end of June. Wirth said, fine, they would adjourn and reconvene again on June 23rd.

As predicted, at the end of June, 1988, the United States was in the throes of a massive heat-wave. A drought was frying the Great Plains. The Mississippi River was drying up. More important, Washington was sweltering. This time, the heat was news.

On this day, when Hansen stepped to the witness table and addressed the committee, his words were carried around the world. The hearing room was jammed to overflowing. Flashbulbs popped. TV lights glared. Hansen spoke.

"The first five months of 1988 are so warm globally that we conclude that 1988 will be the warmest year on record unless there is a remarkable, improbable cooling in the remainder of the year," he said. It was an almost unqualified prediction. Then the other shoe fell. "Global warming has reached a level such that we can ascribe with a high degree of confidence a cause and effect relationship between the greenhouse effect and observed warming."

Emerging from the normal range of variation

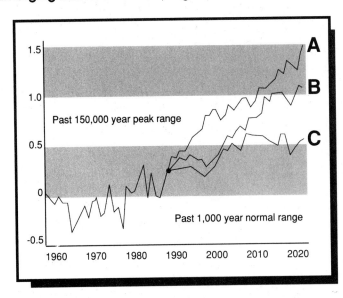

Annual mean surface temperatures have begun to climb above the average range for the past 1,000 years and may soon rise above even the warmest interglacial periods of 6,000 and 120,000 years ago. Scenario A assumes a continuation of present growth in emissions of greenhouse gases. Scenario B would be realized if there were no increase in rates after 1990. Scenario C assumes drastic reductions in emissions between 1990 and the year 2000.

Source: Hansen, 1988.

Hansen concluded, "It is not possible to blame a specific heat-wave/drought on the greenhouse effect. However, there is evidence that the greenhouse effect increases the likelihood of such events—our climate model simulations for the late 1980s and 1990s indicate a tendency for an increase of heat-wave/drought situations in the Southeast and Midwest United States." His detailed analyses were lost to most of the reporters in the room. The headlines of the evening editions screamed: "THE HEAT IS ON!," "GLOBAL WARMING HAS BEGUN," and "TOP SCIENTIST: EARTH OVERHEATING."

The flush of sudden realization lasted through the summer and into the fall of 1988. Stephen Schneider, a climate modeler at the National Center for Atmospheric Research, said he could tell that public interest had reached a new plateau by the sudden increase in the number of phone calls he received daily. He graphed his calls against increasing atmospheric carbon and called it, tongue in cheek, the Schneider Index.

Yet, in public opinion as in physics, for every action there is an equal and opposite reaction. Hansen's statements generated a backlash among the scientific community that challenged his 99 percent confidence and criticized his methods. "They (NASA) have been coupling their atmospheric model to a pretty hokey ocean (model)," said Schneider, "we all have. But you have to have less confidence because of that."

Hansen's staff worked out their climate model on a 1975-vintage "super"-computer. Even to get access to that, the group had to book time late at night and on weekends. Given the resources presently at the disposal of most climatologists, it will certainly be decades before global climate models become sophisticated enough to accurately predict changes in ocean and air currents and to confirm what most people—including most of the scientific skeptics—strongly suspect is going on. The problem is, we don't have decades to study the problem before we must decide how to react.

There is a risk of waiting for more data, of not trusting to intuition, of not changing until more evidence is in. The risk is that we will have to adapt to a much larger amount of warming if we wait, and that the costs of successfully adapting—economic, political, social and other—will have gone up significantly in the interim.

The danger is that after the flurry of press attention to Hansen's statements, and then the flurry of charges and countercharges that consumed much of the following year and succeeded in drawing attention to the uncertainties, that the press will lose interest and go on to some other story, politicians will consider it a matter of scientific debate, and the public will shrug its shoulders and tune out.

Complacency is the greatest obstacle. It takes decades for the climate to change once greenhouse gases have been altered. It takes centuries or millennia before the process can be reversed and the Earth can start to cool once more. We can suddenly decide to stop our warming ways but we cannot suddenly stop the warming or return to the cooler planet that we knew. Time is not on our side. The longer we wait, the longer and more difficult will be the Earth's recovery.

There is only one message in this book. That message is that we should be frightened for the survival of life on Earth. Our physical strength as a species may be exceeding our ability to think collectively and do the right thing. We must begin to communicate among ourselves more effectively. We have a problem that is serious, immediate, growing quickly, and is potentially devastating. We have uncertainties about the pace and distribution of the warming, but we already know enough to justify swift response.

We need to wake up!

Chapter Three

RUNAWAY!

The speed of the changes now under way holds a risk of greater magnitude than most people imagine possible. Over the last 18,000 years, the Earth's temperature has gradually increased. The Earth accommodated this change smoothly, because there was sufficient time for the deep ocean waters to reach an equilibrium with warmer waters closer to the surface. Carbon was removed from the atmosphere by the ability of the surface waters—and the tiny plankton—to absorb it from the air and eventually send it to the oceans' depths. Because of the sudden rise of temperature in the course of a mere 100 years, equal to the rise over 18,000 previous years, and with a loss of plankton due to ocean pollution and ozone depletion, the carbon may not be removed as quickly, and the atmospheric concentration of carbon dioxide may increase far beyond many current predictions. The atmosphere is also experiencing a higher concentration of methane than it has reached in the past 100 million years. What may happen because of these very large and very rapid changes is frightening to contemplate. The net effect could be a runaway greenhouse reaction.

Genesis of Earth's Atmosphere

The dozen most common elements present at the formation of the planets were, in order of decreasing abundance, hydrogen (90%), helium (9%), oxygen, neon, nitrogen, carbon, silicon, magnesium, iron, sulfur, argon, and aluminum. Hydrogen and helium comprised 99% of the available atoms and the remaining 10 elements, along with atoms of other types in trace quantities, made up the remaining one percent.

Of the 12 major types, four—aluminum, iron, magnesium and silicon—only form combinations that are solid and cannot contribute to an atmosphere. They became the core of the planets. Three—argon, helium, and neon—are noble gases which cannot combine with other elements.

With a gradual cooling, the remaining five elements—carbon, hydrogen, nitrogen, oxygen, and sulfur—combined to form molecules that formed the planets' outer crusts and primitive atmospheres. Oxygen and hydrogen combined to form water. Nitrogen and hydrogen combined to form ammonia. Carbon combined with hydrogen to form methane. Sulfur combined with hydrogen to form hydrogen sulfide. Dominating all of the planets' atmospheres were molecules of hydrogen. Because of the temperatures of planets in their beginning stages, these molecules were all gases. All the planets in our solar system may have had similar atmospheres at similar times in their evolution.

Mercury, closest to the sun, quickly lost its hydrogen and helium. The warmth of the sun and the planet's relatively small mass allowed the lightest elements to escape Mercury's weak gravitational pull. Mercury was left with silicon, magnesium, iron, aluminum, and some less common elements.

Jupiter, Saturn, Uranus, and Neptune, which were large enough to develop huge gravitational pulls, all have hydrogen-helium atmospheres. The remaining inner planets, Venus, Earth, and Mars (and Titan, the largest satellite of Saturn), lost much of their hydrogen, helium, neon, and argon to space, but had sufficient gravitational pull to hold a large fraction of these gases close to their surfaces.

These planets' developing atmospheres were composed of ammonia, methane, water vapor, and hydrogen sulfide.

Earth's atmosphere progressed to an abundance of oxygen through the process of *photolysis*, the breaking apart of water molecules by exposure to ultraviolet light. As water vapor rose into the upper atmosphere, it broke apart into atoms of hydrogen and oxygen. The Earth's gravitational field could not retain all the hydrogen but did hold most of the oxygen.

Free oxygen tends to steal hydrogen atoms from other gases and to re-form water molecules. In stealing from ammonia, the oxygen left free nitrogen. From methane, it created carbon dioxide. From hydrogen sulfide, it created sulfur dioxide.

As the planet cooled, the water vapor condensed into a liquid, forming oceans. The carbon dioxide and sulfur dioxide dissolved into the oceans and were deposited in the Earth's crust. The less abundant sulfur was largely scrubbed from the atmosphere in this way, leaving Earth, after many millions of years, with an atmosphere of nitrogen, oxygen, carbon dioxide, and water vapor. It is possible that the same processes took place on Mars and Venus, which at one time may have had limited oxygen atmospheres and been covered with liquid water.

On Earth, life began when the surface was cooling, oceans were forming, and the planet's atmosphere was changing from ammonia, methane, water vapor, and hydrogen sulfide to oxygen, nitrogen, carbon, and sulfur dioxides. Once life appeared, the atmosphere of Earth changed dramatically because the planet developed a method of producing oxygen that was much more efficient than photolysis. The new way, *photosynthesis*, used visible sunlight, rather than ultraviolet light, to transform carbon dioxide into free oxygen and organic compounds. The carbon was absorbed into the living material of plants and became plant or animal tissue (and eventually fossil fuels). The oxygen was exhaled to the atmosphere to gather more carbon.

About 3 billion years ago, the first organisms capable of photosynthesis, one-celled cyanobacteria, began their work to transform the Earth's atmosphere from predominantly carbon dioxide to predominantly oxygen. It took some 2 billion years to accomplish the process. With an increasing abundance of oxygen came a profusion of lifeforms. About 1.4 billion years ago, cyanobacteria began to be replaced by more efficient photosynthesizers, the green algae. As oxygen became more abundant, living creatures were able to thrive in oxygen and to develop muscles, hard skins, and sophisticated organs. When our atmospheric oxygenation was nearing completion, some 600-650 million years ago, Earth underwent an explosion of life.

Three major factors contributed to this explosion: the availability of oxygen, carbon, and water as building blocks of living tissue; the co-development of photosynthesizing and oxygen-breathing lifeforms to ensure a sustainable chemical balance, living on the daily income of sunlight; and finally, the formation of a greenhouse/ozone screen which blocked life-destroying ultraviolet radiation and moderated the climate of the planet.

AXIAL TILT, SOLAR CYCLES, AND ICE AGES

In the life of the Earth, there have been 6 major ice ages. The first was about 2.5 billion years ago, when biological life was only single-celled bacteria. The second was 1 billion years ago, when green algae were also present. There were major ice ages again 700 million years ago, 570 million years ago, 450 million years ago, and 300 million years ago. These events were very likely all caused by slight changes in the radiation of the sun—just a few percent less solar heat arriving at Earth—and the inability of the evolving climate system to moderate the relatively sudden change.

Over the past ten million years, the planet has experienced 10 severe and 40 minor episodes of glaciation, although none approaching the magnitude of the major ice ages. The largest of these smaller ice ages was 1.5 million years ago, when our ancestors lived in caves. The most recent was about 20,000 years ago, when polar ice extended as far south as the Weser Valley in Germany and the Ohio Basin in North America. Sea levels fell as much as 325 feet (100 meters), exposing vast stretches of ocean floor. Eleven million square miles (28 million sq. km.) of Europe, Asia, and North America were under glaciers up to a mile high.

Axial tilt of the Earth

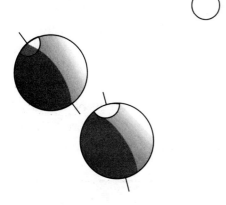

As the Earth's axis tilts toward the sun, more of the North pole is periodically exposed to sunlight, which influences worldwide climate and glaciation.

During a 41,000 year cycle, the inclination of the Earth's axis varies about 1.5° on either side of its present value of 23.5°. This tilt toward or away from the sun affects the latitudinal distribution of solar radiation. It is accepted by most scientists today, after being first suggested by J.A. Adhémar in 1842 and mathematically refined by Scottish millwright James Croll in 1864, that the axial tilt cycle is responsible for most, if not all, of the 10 severe glaciations in the past million years.

There is also a 23,000 year wobble on axis and a slightly egg-shaped irregularity in the planet's orbit which brings us closer to or farther from the sun in 100,000 year cycles. These two seem to work in conjunction, so that the wobble, or precession, is most pronounced when Earth's orbit is most elliptical. While the variation in orbit may increase or decrease sunlight arriving at Earth by only one or two tenths of a percent, the precession can reduce or enlarge the intensity of the solar beam by up to 10 percent for a given season and location.

THE CARBON CYCLE

We know, with a high degree of certainty, that the carbon cycle has corresponded perfectly with many, if not all, glaciations in the past million years. As ice sheets grow, oceans shrink. At the height of glaciation, as much as 10 percent of the oceans' water may become ice. However, since ice cannot hold nearly as much carbonate material as liquid water, carbon dioxide in the atmosphere begins to accumulate. As carbon dioxide accumulates, Earth's temperature rises and the ice melts.

The drop in sea level also bares rock that is rich in phosphate and other minerals. Once exposed to weathering processes, these rock formations leach nutrients into the oceans, which speeds the growth of phytoplankton. With fertilization, the plankton consume more carbon from the air. When they die, the phytoplankton, zooplankton, and marine animals farther up the food chain sink toward the ocean floor. Some of their carbon remains as sediment at the bottom of the ocean and some is dissolved as gas in the ocean water. The fertilization of the oceans, which occurs when sea level is lower, lasts long after sea levels rise again, augmenting the numbers of plankton gathering carbon. As the plankton gradually remove carbon from the atmosphere, the Earth cools, and the ice sheets advance once more.

Warmer and Warmer ...

- More forest fires will send greater amounts of carbon skyward while reducing the rate of carbon removal by forests.
- Faster tree respiration, due to warmer soils, will mean less carbon absorption by forests and could bring about large scale temperate deforestation.
- Organic matter in soil will decompose faster, sending more carbon dioxide and methane into the atmosphere.
- Methane and other greenhouse gases trapped under thawing Arctic tundra will be released to the atmosphere.
- Melting polar ice will reflect less solar radiation to space and increase the rate of warming.
- As oceans become warmer and more saturated with carbon dioxide, they will be less able to absorb carbon dioxide than they have in the past.
- Increased temperatures will cause an increase in irrigation, air conditioning, and refrigeration, causing greater levels of electrical generation, which in turn will generate still more carbon dioxide.

A similar process happens on land. When glaciers advance across the continents, they scour the hillsides, grinding up rock. The crushed rock is later deposited, enriching and remineralizing the soil as the glaciers retreat. Boreal forests return to this fertile ground and grow with renewed vigor, removing carbon from the atmosphere at much faster rates than they had before the glaciation, when soil minerals were depleted. When the high latitude forests reach their full extent, so much carbon is removed from the atmosphere that the Earth begins to cool and the ice returns.

**Global mean temperature and glaciation
from 150,000 years ago to present**

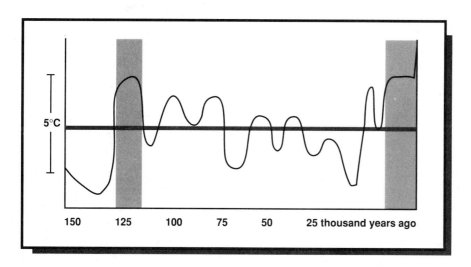

The ice age cycles of the past 150,000 years can be charted by the oxygen content of deep ocean sediments. This record covers just the most recent of eight ice cycles in the past one million years. The shaded areas are the warm interglacials, when polar ice sheets have retreated to their smallest extent.

Source: Broecker, 1987.

When these carbon cycles correspond to changes in the earth's orbit and tilt or other influences on solar radiation patterns, the glaciation is severe. When there is no correspondence, or the correspondence is antagonistic, the glaciation is minor. Carbon cycles do not appear to be governed by astronomical constants and cannot be predicted with any precision. What is predictable is that increases of atmospheric carbon and decreases in ocean plankton or forest cover bring warmer temperatures as surely as summer follows spring.

The rapid buildup of greenhouse gases has now altered our "normal" ice age cycle. In the usual course of events, we should have reached a peak temperature a few thousand years ago and be gradually cooling now. That has not happened. Instead, in the last few centuries we began getting warmer again. The Earth may soon be warmer than it has been between any of the interglacial periods for the past one million years.

THE CYCLES OF THE SUN

Suppose we go back still farther and examine climate for the past 500 million years. During 90 percent of that time, the worldwide average temperature was 14 degrees warmer than today—about 72°F (22°C) compared to 58°F (14°C) today. During the most severe ice ages, the lowest point that average temperature reached was 42°F (6°C), about 16 degrees colder than today.

For the duration of human evolution, we have been blessed with a cooler, more temperate Earth—ideal conditions for mammals such as ourselves. At no time in the past 1 million years have we had to experience the extremes of heat and cold which the Earth had earlier.

The sun is not getting cooler as it ages, however. It is heating up. How then could the Earth's climate have been so much warmer in earlier ages? One reason is that the Earth itself is cooling. Four billion years ago, it was a hot ball of magma. It is now a few hundred degrees cooler and getting cooler still. Another reason can be found in the changing thermonuclear chemistry of the sun. At times the sun is relatively warmer and at other times cooler.

Still, if the sun emitted 25 to 30 percent less energy 4 billion years ago, the surface of the Earth should have been covered with ice and snow. That the Earth was warm at that time is known as the "faint early sun paradox." A solution to the paradox was first posed by astronomers Carl Sagan and George Mullen in 1972, after studying the climate of Mars. Sagan and Mullen proposed that the early Earth was warmed by a "supergreenhouse" of volcanic origin. Eventually the carbon dioxide and methane released by volcanoes was deposited in carbonate rocks and the Earth's temperature cooled, even as the sun's output increased.

THE RUNAWAY REACTION

The seasonality of the climate of 18,000 years ago was similar to today. Between 15,000 and 9,000 years ago, the earth-sun distance decreased and the northern axial tilt increased, causing continental ice sheets to retreat and the oceans to warm up. About 9,000 years ago, the average solar radiation in the Northern Hemisphere was about 8 percent higher in July and 8 percent lower in January than it is today. Over the past nine millennia, this seasonal fluctuation gradually steadied.

What will be the effect of increasing axial tilt in combination with greenhouse warming? As we move past the point of equilibrium we are now experiencing and tilt back from the sun, the Southern Hemisphere will experience greater warming while the Northern Hemisphere may experience wider seasonal swings. Axial tilt is a phenomenon that occurs over thousands of years, however, giving plants, animals, and humans, ample time to react to changing conditions. The greenhouse crisis of the next century will not be nearly as forgiving.

Temperatures changes during the last ice age in degrees Celsius difference from today

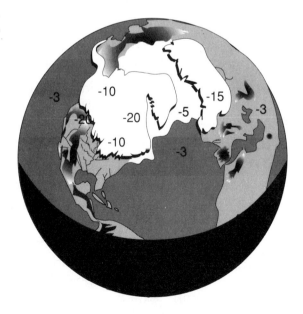

Sources: Hansen, 1987; Imbrie and Imbrie, 1979.

When we look back over the processes that made our climate—our planet's formation from hot stellar gases, the billion-year progression of atmospheric chemistry, the co-evolution of carbon-oxygen biological and geophysical life-cycles, and finally, the arrival of human beings—we have to consider ourselves pretty lucky. Anywhere along the way something might have gone wrong and our planet could have become as lifeless as Mars or Venus.

It is hard to imagine that the geophysical evolution accomplished over 3 billion years could become undone in a single century. Yet the rate at which we humans have been combining stored fossil hydrogen and carbon from the Earth's crust with the oxygen in our air has been, in planetary terms, very fast and steadily increasing. Through the power of our intelligence, we have compressed processes that used to take millions of years into mere decades. We are consuming not only all of the world's present forests, but all of the forests of the world that existed in ancient ages and were trapped by geological processes and transmuted into oil, gas, and coal.

While our alteration of Earth's carbon cycle is extensive, the proportionate effects we have had on other, more powerful atmospheric gases is even greater. We are not in danger of suffocating from a lack of oxygen. What we are endangering is something we have always taken for granted—the temperature.

We have been blissfully oblivious to the fragility of the Earth's climate systems. We have been overawed by their size and complexity. We must now begin to recognize that there is only a very delicate balance that maintains climate, that we have become able to alter that balance by the power of our industry, and that we are now altering it in a haphazard way, very quickly.

Chaos

One of the first persons to put a computer to work on the science of climatology was Edward Lorenz at the Massachusetts Institute of Technology. To better understand the natural processes at work, in 1960 Lorenz attempted to make the university's Royal McBee computer print out artificial weather. For more than a year, Lorenz's programs provided computer-generated patterns of wind, heat, and precipitation that were very sensible, but very predictable. The computer just modified and returned whatever Lorenz typed in, following preset rules in an unchanging sequence. The model could become as complex as Lorenz cared to make it, but it was not realistic. True weather is unpredictable.

Then one day in the winter of 1961, Lorenz took a fortuitous shortcut. Instead of starting the whole computer run from the top, he typed in the numbers from the middle of an earlier run, left the computer to replot the graph, and walked down the hall for a coffee break. As he sipped his coffee, a new branch of mathematics was invented.

Lorenz's first chaotic weather pattern

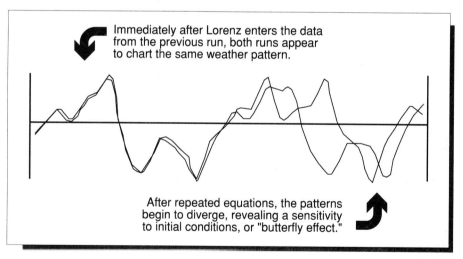

Source: Gleick, 1987.

When Lorenz walked back to his office and looked at the printout, what he saw was something completely different than what he had expected. Instead of the same weather pattern as before, the computer had created something quite new. The repeat pattern started at the same point and followed the previous pattern closely for a short time, but then began to diverge. It continued to diverge until all resemblance to the original sequence disappeared. Lorenz could have assumed something was wrong with the computer, or his program, but instead he guessed, correctly, that he had stumbled onto something profound. He had discovered a mathematical key to chaos.

Lorenz's diverging pattern was caused by the significant difference between the six-decimal numbers used by his computer (.506127) and the rounded-off three-decimal numbers appearing on the printout he had rekeyed (.506). When he typed in the shorter number, Lorenz had thought that one part in ten thousand would be inconsequential. After all, in numbers referring to windspeed, a ten-thousandth part difference represents only an imperceptible puff of wind, not an entire weather system. But as the difference propagated itself in equation after equation, the entire weather of the Earth changed. Lorenz named the phenomenon the "butterfly effect"—because it now seemed that a butterfly stirring the springtime air in Peking could transform the course of summer storms in New York.

Lorenz reasoned that sensitivity to initial conditions was no accident, but is necessary to all natural systems. The influence of small perturbations is what endows larger patterns with such rich variety. It is what gives weather its unpredictability.

Lorenz had made a conceptual breakthrough, but he wasn't finished. He next decided to pursue the mathematics of fluid systems by modeling the chaos of convection. In a closed container, such as a pot of coffee, the temperature difference between the hot bottom and the cool top controls the flow of fluid—the motion of the boiling water. As the fluid below becomes hot, it expands. As it expands, it loses density, becomes lighter, and rises. In a perfectly shaped container, a cylindrical roll would develop, with hot

The Lorenz Attractor

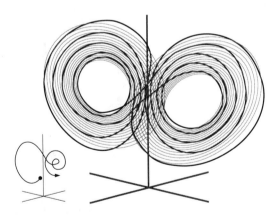

Slight variations which continuously diverge but never intersect or repeat are represented in Lorenz's graph of the turbulence of a convecting fluid over time. The crossover of one wing to the other represents a sudden change, such as a reversal in the spin of the fluid.

Source: Gleick, 1987.

fluid rising on one side and cooler fluid sinking on the other. But tiny imperfections, such as the shape of the pot or the weight of the grains of coffee, make a cylindrical roll more difficult. They impede the momentum of the hot fluid. They create eddies and counterflows.

Lorenz discovered that even in the chaos of convecting fluids there were larger patterns that seemed to establish themselves. When he graphed the convective flow over time, he discovered a pattern in the way that the trajectory never exactly repeated itself, but instead looped around in a constant state of variation. Ironically, the graph resembled a butterfly's wings.

As long as the convective system remained within the confines of some parameters, such as the amount of heat applied, the pattern of turbulent flow became a regular double spiral in three dimensions. No arrangement of water molecules ever recurred, but the shape of the sequence of arrangements was orderly.

Scientists who later improved on Lorenz's work discovered that if the parameters were enlarged to introduce greater disorder, such as by adding more heat, the butterfly pattern became much wilder, flipping into larger, smaller, or entirely different shapes before finding new equilibrium.

More than two hundred years ago the Dutch physicist, Christian Huygens, observed a set of pendulum clocks swinging in perfect synchronization and realized that every clock's mechanical springs, friction, and weights could not be that perfectly equal. Huygens found that the clocks were coordinated by vibrations transmitted through the wood in the walls and floor. Huygen's pendulum phenomenon is now known to physicists as entrainment, or "mode locking."

Destruction of the butterfly

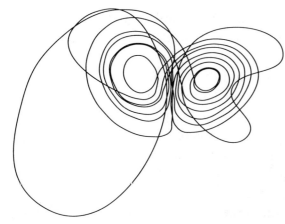

Forcing the parameters, such as by adding heat to a convecting fluid, can change the dynamics of the sequence, flipping it into wild leaps of disequilibrium before reestablishing a pattern. Linear graphs show gradual increases in global temperature and precipitation over several decades, but this computer generated distortion of the Lorenz attractor portrays the sudden disruptions of seasonal climate patterns that could come with global warming.

Mode locking explains why the moon faces the earth, why a radio receiver can home in on a weak signal even when there are variations in the transmitting frequency, and how crickets gather together by the thousands and chirp in perfect unison.

The processes of weather and climate are both mode locked and chaotic. By sharply elevating one element—the heat of the whole system—in a short amount of time, the mode lock can be broken and the pattern made more turbulent.

The Earth's climate, upon which life depends, is a chaotic system, but it also follows the rules of chaos. Global warming could change the global climate system from one that appears to be stable but is really a constant interplay of random events—locked in a mode of orderly disorder—into something that is so disorderly it can no longer be recognized as a organized system.

WHAT IF ...

If climatologists have nightmares, they may be about possibilities that are almost too horrendous to contemplate during waking hours. What if ozone depletion increases ultraviolet radiation to the point of killing ocean plankton? Or what if pollution of the oceans kills the plankton? What if global warming, increased ultraviolet radiation, acid rain, or toxic pollutants begin to shrivel fields and woodlands, in the process releasing still more gigatons of carbon to the atmosphere and setting off a chain reaction? After most of the plankton and the trees are dead, how will the atmosphere be cleansed? What if polar warming releases all the methane now trapped under permafrost and tundra? What if the oceans reach a temperature at which carbon dioxide will no longer be absorbed? What if the West Antarctic Ice Sheet melts? Would the increase in sea level change the Earth's center of gravity? Could the North and South poles shift?

The enormity of the crisis we have so recently discovered offers no reassurances. The scope and speed of the climate changes, our lack of information about coupled systems, and our limited ability to influence human behavior all make it probable that more large surprises lie ahead.

When Venus spun away from the sun almost 5 billion years ago, it was essentially the same size and composition as Earth. Today Earth is a blue water world with an oxygen atmosphere and abundant life. Venus is a lifeless, bone-dry rock shrouded in dense clouds of sulfuric acid. The surface of Venus is hot enough to melt lead.

Without its thick clouds, Venus would be approximately the same temperature as Earth. However, because its carbon dioxide atmosphere traps infrared radiation 100 times more efficiently than our atmosphere, Venus is 750°F hotter. Could Venus have once been like Earth, a blue planet covered by oceans? Could Venus's atmosphere, eons ago, have begun a carbon dioxide exchange cycle that got away and changed that planet forever? Could a runaway greenhouse effect have consumed its oxygen, evaporated its oceans, and turned its surface into a hellish oven incapable of sustaining life in any form?

The Goldilocks Phenomenon
Average temperatures of the planets

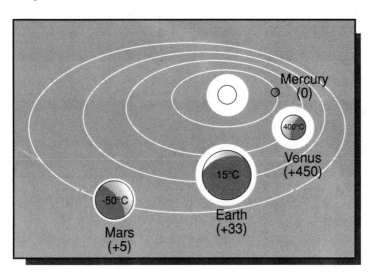

Greenhouse contributions in degrees Celsius are shown in parentheses. Mercury, closest to the sun, has no atmosphere and is extremely hot. Mars, farthest of the 4 planets, has a weak greenhouse effect and is very cold. Venus would be the same temperature as Mars without its thick atmosphere, but it is 450°C warmer because of a strong greenhouse effect. Earth, with its moderate atmosphere, is warmer than Mars and cooler than Venus—a perfect temperature for life.

As we look around the worlds within our solar system, we see no other life-bearing planets. This is the only one. Many of the processes which brought us to our present circumstances are processes that are capable of being reversed or overcome by new processes. We are taking enormous risks by tampering with our spaceship.

And yet we neither fully understand nor appear to consider these risks serious enough to give us pause.

We are falling, falling, falling.

Is our plunge into a hellish world of fire, flood, death, and mass extinction something that *must be* or something that *might be?* Might be. But *will be*, unless ...

Chapter Four

THE RISING TIDE

For as long as men have put to sea in sailing ships, a trap for the unlucky sea captain has lain west of the Canary Islands, astride the Atlantic passage to the New World. There, in the middle of the ocean, a thousand miles from land, the sea surface suddenly changes color. Derelict ships, their wooden masts bleached by the sun, lie still, as if at anchor. Though the ocean bottom in that spot is as deep as the Himalayas are high, there are grasses and weeds which peer up from just below the surface, giving the place the look and smell of an inland marsh. This deep ocean swamp, where wind does not blow and oar does not row, is a graveyard for the unwary. It is the Sargasso Sea.

It was the sun that created the Sargasso. When the sun's heat reaches Earth's surface, the warm air rises until it gets close enough to space to give up its heat, and then it sinks earthward once again. The movement of the hot air up, cool air down, and the rotation of the planet combine to make "cells" of circulation which whirl wind and ocean in familiar currents that are as old as recorded history.

In the North Atlantic, the prevailing winds are the Westerlies, driving hard and cold from Newfoundland to Norway. Farther south, the prevailing winds are the Northeast Trades, blowing warm and steady from the Mediterranean to the Caribbean. Because of the shape of the continents and the direction of the Earth's rotation, these winds have created an elliptical gyre of ocean current which constantly rotates clockwise. From Cape Hatteras to Nova Scotia, the Atlantic current, 50 to 100 miles wide and a mile deep, is called the Gulf Stream. Between Iceland and Scotland, it becomes the North Atlantic Drift. As it returns Southwest from the Azores to the Windward Islands in the West Indies, it is the North Equatorial Current. In the center of the gyre, where the water is as still as air in the eye of a hurricane, is the Sargasso Sea.

Atmosphere and oceans are joined together at the sea surface. The joining is not a solid weld, like two pieces of metal, but a rhythmic exchange of caresses. They exchange energy. The wind blows on the sea and generates waves. The waves drag on the atmosphere's momentum. There are chemical exchanges, conduction and convection.

Water is much heavier than air and its heat storage capacity is much greater. Because of that, the oceans are slow to respond to changes in air temperature. The ocean retains a long-term memory of past weather and climate.

Climate as we know it is largely controlled by the oceans' memory. Northern Europe is warmer than Eastern Canada because as weather crosses the Atlantic Ocean it picks up heat from the Gulf Stream. That heat developed during the preceding summer months and from the time the water in the current was at the equator. The Bahamas are cooled in much the same way that Northern Europe is warmed. Ocean memory helps to moderate the weather.

Ocean memory is vast. It remembers the "Little Ice Age" from 1645 to 1715, when a drastic drop in sunspot activity (the Maunder Sunspot Minimum) coincided with a half-century drop in global temperature of about 0.75°F, causing the canals of Venice to freeze over, French agriculture to collapse, and Louis XIV to install parquet over the marble floors of his palace at Versailles. The ocean is still cooling from those winters in the depths of its deepest abyss. It remembers the Summer of '88, and will keep that memory for centuries to come. The ocean also remembers air chemistry, because of the molecular exchanges between air and water. Ocean memory contains a clear record of increasing carbon dioxide, freons, toxic waste discharges, radioactivity, oil tanker accidents, and other human influences.

Ocean convection patterns

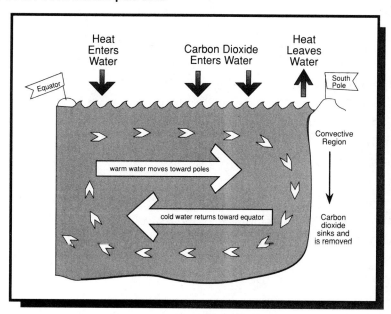

Differences in temperature create circulation patterns that carry surface waters poleward and return deeper ocean waters toward the Equator. Changes in temperature differential which may accompany global warming could reduce the strength of these convection currents and lessen the ocean's ability to absorb and deposit out carbon dioxide.

Source: Wunch, 1988.

The ocean currents are enormous conveyor belts, hauling greenhouse gases, temperature, and ocean life up and down and around the globe. We know from observing these currents that atmospheric heating is affecting the ocean. Sea level is rising. The ocean is warming. We don't yet know what influence the greenhouse effect may have on ocean currents. Atmospheric models suggest that the most profound warming will be near the poles. An increase of heat there could bring a reduction of wind force. A reduction of the winds at the poles would mean that less heat would be absorbed by the oceans at high latitudes. This change would reduce the strength of parts of the conveyor belt.

As the temperature difference between the equator and the poles narrows, the prevailing Westerlies may weaken. Weaker winds would mean a weaker Gulf Stream. Less warm water reaching Europe could mean that the European climate will cool. But there is another possibility.

As the Atlantic gyre weakens and prevailing winds slow, the Sargasso Sea could grow larger, or migrate northward. Westerly winds blowing over the warm Sargasso could make Europe much warmer.

OCEAN SEASONING

Polar winds are not the only driving force at work. Salty water is heavier and has a lower freezing point than fresh water. Because evaporation exceeds precipitation in the Atlantic Ocean, the waters between Nova Scotia and Northern Europe are salty and tend to increase in density as they cool. The denser waters sink more rapidly, drawing the ocean currents into a rolling motion. The reverse situation exists in the Pacific, where precipitation is greater and the salt content is lower.

Ten to eleven thousand years ago, the chemical composition of the North Atlantic changed radically for reasons that are still uncertain. This period is called the Younger Dryas, after a tiny flower that grew in the shadow of the retreating glaciers. During the Younger Dryas, Northern Europe and the Maritime Provinces of Canada suddenly reverted to the climate of the preceding ice age. The onset was rapid—less than a century from warm temperatures to deep ice. But then, just as suddenly, the ice vanished.

One theory to explain the Younger Dryas is that of a large release of glacial meltwater from inland Canada to the North Atlantic. Since salty water does not form ice easily, continental areas adjoining the North Atlantic are kept warm by the circulating salty currents. If the

North Atlantic suddenly became less salty, ice would have formed on the ocean surface, the North Atlantic currents would have slowed, and the ice-chilled winds would have picked up. These effects could explain a rapid drop in the temperature of Northern Europe.

Wallace Broecker, a geochemist at Columbia University, has suggested that the Younger Dryas may have been initiated when a melting ice sheet that had been functioning as a wall channeling meltwater into the Mississippi River and the Gulf of Mexico suddenly broke apart and allowed melting ice water to flow by way of the St. Lawrence River directly into the North Atlantic.

This theory of the Younger Dryas lends credence to the mounting evidence that relatively small changes in the chemical balance of the atmosphere and the oceans can trigger large changes in sea ice cover, ocean currents, precipitation, and climate. As climatologist Stephen Schneider has said, "the Younger Dryas does dramatically remind us that the climate system is a very complex set of interacting subcomponents whose workings are far from being fully understood. Quite simply, the bottom line is that if we disturb the climatic environment by as much as nature has from the last ice age to the present interglacial, and we do it some ten to fifty times faster, then nasty surprises, such as a radical and potentially catastrophic shift in ocean currents, are plausible. And the faster we alter the climate, the greater the likelihood of surprises."

SEA LEVEL RISE

The rise of sea level is the most easily predicted consequence of global warming, and potentially the most devastating. Warmer temperatures will bring more evaporation and more precipitation worldwide. As temperatures rise, the waters of the Earth will expand. Polar icecaps and mountain glaciers will melt at faster rates.

The one degree increase in temperature over the past century contributed to a 4 to 8 inch (10 to 20 cm) rise in mean sea level. Considering only the changes now underway in the atmosphere, and not considering new or accelerated emissions, unexpected volcanic activity, or other factors, the rise of the world's oceans over the next century will average one inch per year. The rise in mean sea level could be from as little as 2 feet to as great as 14 feet by the year 2100.

These figures take account of an average sea level world-wide, but actually sea level is not uniform. For example, the Atlantic and Gulf coastal waters of North and Central America will rise 6 to 9 inches more than the Pacific waters because of the continental shelf and ongoing geological processes.

Because most coasts are sinking (and a few are rising), lowlands in California and the Gulf Coast may experience a relative sea level rise of 3 to 6 feet (1 to 2 meters) per century. A number of islands close to the Equator may vanish beneath the waves. Meanwhile, in Hawaii, at Iceland, and along the coast of Alaska, the shoreline may become more elevated because of volcanism, mountain uplifting, and isostatic rebound (the "bounce back" of land to the elevation it had before being depressed by the weight of glaciers).

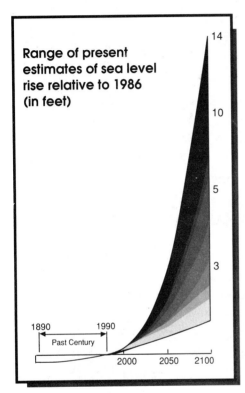

Source: Smith and Tirpak, 1988.

These estimates do not account for catastrophic events, such as the separation of the West Antarctic ice sheet. If that large mass of ice—three miles thick and comprising nearly three-quarters of the world's fresh water supply—should decay enough to slide into the ocean, Earth's oceans could rise 23 feet (7 meters) in a single year.

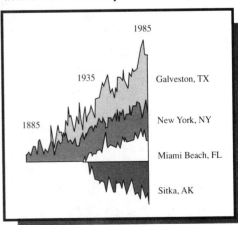

Source: Smith and Tirpak, 1988.

Because of the circulation pattern of ocean currents near Antarctica, it is possible that the Southern Hemisphere may experience less dramatic climate changes than the Northern Hemisphere. If this occurs, the warming of southern ice may be less than predicted and sea-level rise will not be as severe.

Changing Shorelines

The physical consequences of sea-level rise include shoreline retreat, temporary flooding, and salt intrusion. In the United States, a rise of 4 to 7 feet would permanently inundate major portions of Louisiana and Florida, destroying many refineries, grain terminals, shipyards, beach resorts, and residences. Eighty percent of all wetlands, which harbor most coastal wildlife, would be inundated. A 4-foot rise would also flood up to 10,000 square miles of inland areas.

Effects of Sea Level Rise

- Severe and frequent storm damage
- Flooding and disappearance of wetlands and lowlands
- Coastal erosion
- Loss of beaches, low islands and spits
- More severe and frequent storm flooding
- Loss of coastal structures, both natural and man-made
- Wildlife extinctions
- Increased salinity of rivers, bays and aquifers
- Landward migration of threatened populations

World climate depends largely on circulation patterns by which the atmosphere and the oceans transport heat from warm to cold regions. An important effect of sudden warming will be shifts in annual and seasonal storm patterns, with some areas gaining and others losing. Because hurricanes and typhoons require an ocean temperature of 80°F or warmer, we can expect powerful storms to form at higher latitudes and during a greater part of the year. Where hurricanes now form, we can expect to see several hurricanes forming simultaneously, or in rapid sequence, sometimes battering the same shore area in successive waves. *It is during these short and very dramatic events that most of the damage to coastal areas will be accomplished.* Because warmer water pumps more evaporative energy into tropical depressions, the storms of the future will become much more intense as the oceans warm. According to meteorology professor Kerry Emanuel at the Massachusetts Institute of Technology, a two degree increase in ocean temperature will raise average hurricane intensity by 40 percent. With a two degree global warming already a foregone conclusion, it is a safe bet that the hurricanes of the next century are going to break all previous records.

DIKES AND SEA-WALLS

Without carefully constructed dikes and sand dunes, half the Netherlands would be under water. Over the centuries, some 370 miles of dikes and dunes have been built to reclaim valuable land from the shallow river deltas of the Meuse, Rhine, and Scheldt rivers as they flow into the Atlantic Ocean.

In 1953, a storm surge destroyed a hundred miles of dikes, inundating 600 square miles of land and drowning more than 1,800 people. In response, the Dutch government put together a Delta Plan to better protect the country from storms. The plan anticipated worst-case storm surges up to those occurring once in 10,000 years. The dikes that were built to protect against these major events were very expensive. A single 2.4 mile dike completed in 1986 cost $3.2 billion. However, the rapid rise of the ocean will now increase the frequency and power of storm surges far beyond what was anticipated in the Delta Plan. Storms that were to have occurred every 10,000 years may now occur every century. The Netherlands will have to invest tens of billions more dollars in dike improvements, just to keep pace.

North Carolina in the 21st century

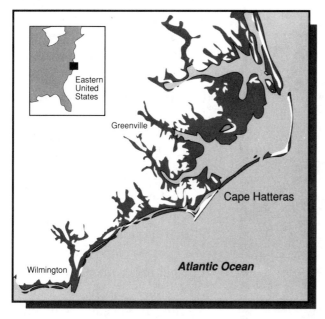

The dark area between the white landmass and the Atlantic Ocean represents lands that would be inundated with a 5-foot rise in mean sea level and no protective measures.

Source: Wilms, 1988.

As difficult as this economic burden may be on the Dutch people, it is small compared to what the United States, with 19,000 miles of coastline, will have to spend. At direct risk are cities, residences, parks and wildlife sanctuaries in Maine, New Hampshire, Massachusetts (including Cape Cod, Martha's Vineyard, and Nantucket), Rhode Island (and Block Island), Connecticut, New York (and Long Island), New Jersey, Delaware, Maryland (and Assateague Island), Virginia (including much of the Chesapeake Bay), North Carolina (and its Outer Banks), South Carolina (including Hilton Head), Georgia, the Everglades and coastal spits of Florida, the bayous of Louisiana (and New Orleans), the Texas Gulf Coast, California (especially the San Francisco Bay), Oregon and Washington.

While several major cities such as Boston, Charleston, Miami, and New York have extensive commercial development less than 4 feet above the ocean, it is doubtful that these areas will be flooded. The rate of sea-level rise is gradual enough and predictable enough, and the land is valued highly enough, that coastal engineering structures will almost certainly be erected to hold back the sea. However, areas with fewer resources than these major metropolitan areas may find the costs of building barriers to be greater than the value of the land being saved. Thousands of miles of coastline will simply have to be abandoned to the ocean.

Bradenton Beach, Florida, in the 21st century

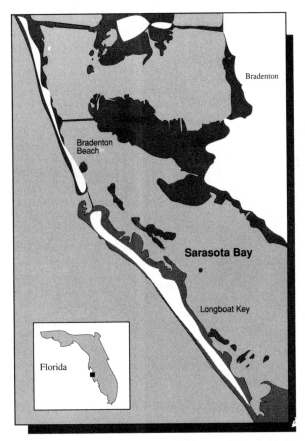

The dark area between the white landmass and the Gulf of Mexico represents lands that would be inundated with a 5-foot rise in mean sea level and no protective measures.

Source: U.S. Geodetic Survey.

Low-lying river deltas with human populations face a double challenge. As sea level rises, storms will bring ocean waves crashing over them. As delta residents pump fresh groundwater to drink or use for irrigation, they may find increasing salinity, because rising oceans will cause sea water to infiltrate into freshwater aquifers.

Deltas are formed from the sand and silt carried by rivers. In normal delta-building processes, rivers deposit their sediments as the channels widen out and the waters slow down just before reaching the ocean. Sediments accumulate to form marshes and swamps, then islands and spits, and then dry land. Local tectonic effects (earth tremors) and compaction cause delta land to normally subside several inches per year unless additional sediments are added through regular river flooding.

When people begin to inhabit the deltas, they dig channels and erect levees to prevent the regular floods. Delta residents pump groundwater to drink. The more they pump groundwater, divert rivers and compact the soil, the more the land subsides. In Bangkok, Thailand, for instance, net subsidence is now five inches per year. Even without global warming, inhabited deltas tend to sink below sea level. Global warming and the accompanying sea level rise will greatly accelerate this process.

San Francisco Bay in the 21st century

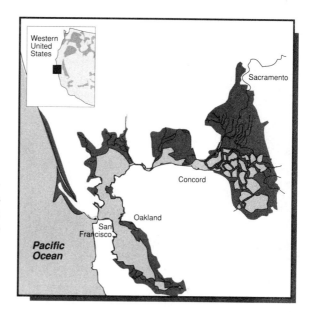

The dark area between the white landmass and the Pacific Ocean represents lands that would be inundated with a 5-foot rise in mean sea level and no protective measures.

Source: U.S. Geodetic Survey.

Tokyo Bay, Japan, during the Jomon transgression

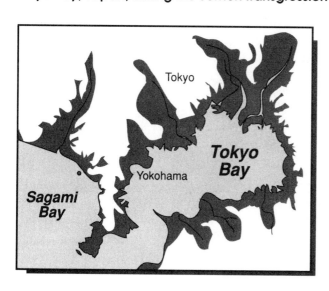

From 15,000 to 6,000 years ago, sea level off Japan rose very rapidly as ice in the Northern Pacific rim melted. By examining raised sea-caves, fossil shell beds, and elevated coral reefs, scientists are able to estimate the shoreline during this period. While future sea level rise may not follow the same pattern, the Jomon transgression is useful to illustrate the extent of shoreline change which can be attributed to a 5°C rise in average temperature. The dark area between the white landmass and the Pacific Ocean represents lands that were inundated with a 5-foot rise in mean sea level.

Source: Machita, 1975.

Channelization

Where humans channelize rivers, sediment either carries past lowlands and is borne out to sea, as it is in the Mississippi Delta, or it is blocked and kept up-river, as it is in the Nile Basin. In either case, no new sediment is being deposited on the deltas, so normal compaction processes cause the landmass to drop relative to sea level. Louisiana, which compounds these other compacting processes by withdrawing underground reservoirs of oil and gas, now loses more land to subsidence and sea level rise than any other area of the world.

If you were planning to buy some real estate near New Orleans, you had better ask how many feet above sea level your prospective purchase stands. Most of Orleans Parish is five feet below sea level. If you were planning to build a home in Key West, consider this. By 2050, many Florida Keys may no longer exist.

In the past we were told the world was shrinking, but we knew it wasn't really, it only seemed that way. Today it really is.

**Future Florida coastlines
with 4.6 meter and 7.6 meter rises in sea level**

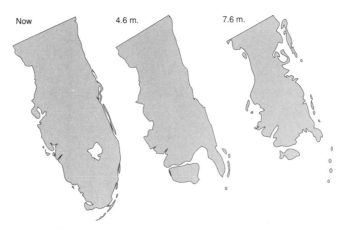

While these scenarios of sea level rise are unlikely to unfold in the next century, they represent what would happen if global warming were to eventually melt or dislodge the Greenland Ice Sheet (middle drawing) and the West Antarctic Ice Sheet (right drawing).

Source: Schneider and Chen, 1980.

BEACHES

Many coastal areas with sufficient elevation to avoid inundation will be threatened less by flooding than by erosion. Just the sea level rise of 2 inches that has already taken place this century has caused serious erosion problems for many coastal areas.

Beach erosion

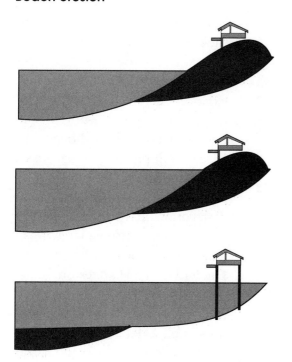

The Bruun Rule. As sea level rises, beach erodes and shoreline retreats. Artificial means, such as "beach nourishment" can supply sand to grade the sloping ocean floor. Without such measures, waves will transport the necessary sand from the upper beach.

Source: Titus, 1986.

Erosion is normal for sandy beaches. The waves drive in harder in the winter, erode the upper beach, and deposit sand offshore. The calmer waves in spring and summer redeposit the sand and restore the beach to its pristine beauty.

When sea level rises suddenly, this balance is disrupted. The waves break farther up on the beach profile. Upper beach sand is deposited farther offshore, meaning that less will return with calmer water until centuries of deposits rebuild the underwater slope. The more the beach retreats, the longer the rebuilding time before the beach will be restored.

Cliffed coasts often have a thin protective beach, which disappears during storms, allowing the waves to attack at the base of the cliff. This process undermines the cliff face, which causes a recession of the cliff. Hard rock cliffs are little affected by even intense wave action, but sandy cliffs can be cut back 10 feet by a single storm. As sea level rises, erodible cliffs will march away from the sea at an unusually rapid pace.

Coastal barrier islands, such as those all along the North American Atlantic and Gulf coasts, will experience beach erosion from increasingly frequent overwash. Many American beach resorts and coastal communities lie on narrow islands and spits (peninsulas with the ocean on one side and a bay or waterway on the other). Rather than erode in place, the overwash experienced during storms will cause these islands and spits to migrate landward. Although this process will protect the barrier islands (some may even increase in elevation), it will wreak great destruction on the seaward side. In this category are some large urban areas, such as Miami Beach, Fort Lauderdale, and Galveston.

In the early 1980s, the Environmental Protection Agency commissioned a detailed study of sea level rise in Galveston and Texas City. EPA's scientists found that a substantial rise in sea level would mean that storm effects which Galveston now experiences on an average of every 100 years (class 3 hurricane), would be experienced every 10 years on average. Even if Galveston were to raise the height of its seawall, which it built after the 1900 hurricane killed 6,000 people, storm surges would enter Galveston Bay through the Bolivar Roads Inlet and flood the city. The seawall would do little to abate the ferocity of the deep ocean swells, because the funnel shape of Galveston Bay would still amplify surge magnitude. A series of large

Palm Beach, Florida, in the 21st century

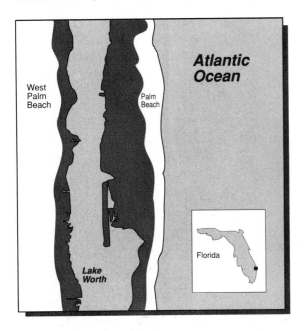

The dark area between the white landmasses and the Intracoastal Waterway represents lands that would be inundated with a 5-foot rise in mean sea level and no protective measures.

Source: U.S. Geodetic Survey.

movable gates, such as are used to protect London, England and Providence, R.I. from storms, were considered but deemed unworkable. In order for gates to protect a bay, there must be no holes in the enclosure. In the case of Galveston, the barrier islands themselves are the holes, because the ocean can go right over them in a hurricane.

Another problem the Galveston study revealed was hazardous waste storage. In Galveston and Texas City, by the middle of the next century, 11 out of 15 hazardous waste sites will be within the 10-year floodplain and two may be completely submerged.

Over 1,000 active hazardous waste landfills and at least that many inactive waste sites in the United States are located in 100-year floodplains. Sea level rise could increase the washout rate of these sites, spreading the wastes and endangering coastal populations for extended periods. In Charleston, S.C., all the hazardous waste landfills in the 100-year floodplain, and some of those on higher ground, will be within the 10-year floodplain by the year 2075.

ESTUARIES

Estuaries are formed where rivers meet the sea. As the rivers widen out, they feel the influence of the ocean's tides. Salt water moves up river at high tide and recedes at low tide.

Estuaries are an important border zone for marine and coastal wildlife, because many marine species depend on freshwater zones to feed and breed and many freshwater species use the changing tides to assist with food gathering and to hide from predators. Sea level rise will move salt water upstream. This will reduce the size of the estuarian habitat and funnel wildlife landward by way of the river.

Over the last 18,000 years, many freshwater rivers like the Susquehanna have evolved estuaries such as the Chesapeake Bay. A decrease in the flow of the river, coupled with a rapid increase in sea level, would cause salt water to encroach farther upstream. A rise of 5 inches over the next 20 years would move salt concentrations in the Delaware River one or two miles upstream. A rise of 3 feet over the next 50 years would threaten Philadelphia's water supply, 20 miles upstream. Because many rivers recharge aquifers, the aquifers would become salty as well.

Many marine species must swim into fresh water during reproduction. When salt water moves farther upstream, so must these fish. But because of global warming, the waters upstream will be shallower, warmer, and more polluted. Without time to adapt, some species will die.

WETLANDS

At the borders of estuaries are the fertile marshlands that are biologically important to the life cycles of a large number of plant and animal species. Wetlands usually keep pace with the gradual rise of sea level between glacial epochs. They collect sediment and produce peat. Swamps and marshes build sediments vertically and expand horizontally into the newly flooded areas. Wetlands are also nature's water purification plants. As floodwaters pass through shallow wetlands, the sediments are deposited and the waters are filtered. This cleansing process allows fish downstream to survive and thrive in times of high water, when soil runoff increases river turbidity (murkiness).

Because marsh vegetation can collect sediments and build upon itself, it has always endured changes in sea level. When increases in sea level on the order of several feet occur over the space of decades, not centuries, the abilities of marshes to gather sediment and grow is lost. The vegetation will drown. Saltwater intrusion and rising water levels may destroy vast areas of flora and fauna, including the Everglades in Florida and the bayous of Louisiana. In many areas, people have built levees and bulkheads just above the marsh. As sea level rises, the wetlands will be squeezed between the sea and the barriers. Removal of these

Wetland migration

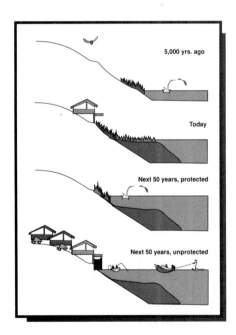

If sea level rises faster that the ability of the marsh to keep pace, wetland area will contract. Construction of bulkheads or other artificial barriers may help adjacent landowners avoid economic damage, but such efforts prevent new marsh from forming.

Source: Titus, 1986.

man-made structures could save 10 to 20 percent of the wetlands from destruction. Without removal, 50 to 80 percent of wetlands will disappear.

Many wetland areas may vanish as soon as the year 2040, when the Gulf of Mexico may surge as far inland as 33 miles. Migration of the wetlands inland will be impossible because of residential development, highways, and other human-erected barriers. With the delicate marsh ecology upset, fish and wildlife industries will decline and the coastal economy will suffer. Communities will lose water supplies and the tax base to pay for replacements at just about the same time.

In coastal aquifers, which are underground rivers, a layer of fresh water floats on top of the heavier salt water. Usually the farther inland you go, the deeper the underground boundary will be between fresh water and sea water. A great many population centers in Louisiana, Florida and elsewhere depend on coastal aquifers for drinking water, but as sea level rises, so will the boundary between fresh water and salt water. The saltwater boundary will penetrate farther inland. Many freshwater wells will become salty. This problem may be accelerated as global warming increases irrigation and other demands. Overpumping of the groundwater will hasten saltwater intrusion.

The Washington D.C. estuary

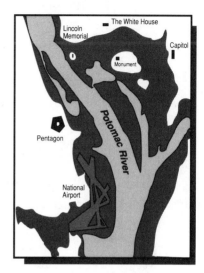

The dark area between the white landmass and the Potomac River represents lands that would be inundated with a 7.6 meter rise in mean sea level and no protective measures.

Source: Schneider and Chen, 1980.

RISE OF SEA LEVEL IN THE THIRD WORLD

If the protection of coastline will be difficult for Americans and Europeans, it will be nearly impossible for people in lesser developed countries. Suriname is a former Dutch colony on the northern coast of South America. For the past decade it has spent nearly twice as much as it has earned in an effort to develop its economy. One of

Suriname's principal creditors is the Netherlands. In Suriname, the per capita income is about $1,200 per year. In the Netherlands, it is $13,000 per year. In a spirit of generosity, the Netherlands, while spending some 6 percent of its gross national product on its own dikes, has lent Suriname more than one billion dollars in order to develop agriculture by erecting dikes on its Atlantic coast. A half-million residents of Suriname are now dependent on those dikes holding back the ocean while they labor to produce the rice, sugar, and fruits that the nation needs to repay its development loans and establish economic stability. To the Surinamese, any rise of the Atlantic Ocean is potentially devastating.

Bangladesh flooding from soil erosion and sea level rise

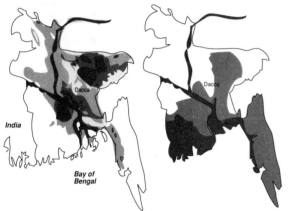

Flooding from rainfall   Flooding from sea level rise

The map at the left shows present-day flooding patterns in Bangladesh, which are attributable to deforestation, soil erosion, and land mismanagement in the Himalayan foothills, as well as to unusually high rainfall. Darker shading represents flooding which is more frequent and severe. The right map shows additional flooding which can be expected as the Bay of Bengal rises and storms carry sea water farther inland. Very little land is left unaffected.

Source: Manolo, 1977.

The recent flooding in Bangladesh is all too familiar. Millions of people in Bangladesh live only a few feet above sea level. Eighty percent of the country is on the Bengal Delta, at the confluence of the Ganges, Brahmaputra, and Meghna rivers, and much of what remains is water. Ocean warming will increase the severity of the monsoons that in the past have carried storm surges 125 miles inland and flooded one-third of the countryside. Like Suriname, Bangladesh is not nearly as well equipped as the Netherlands, the United States, or other countries to devote vast sums of money to the construction or elevation of dikes and sea walls.

In 1960, floods in Bangladesh killed 10,000 people. In 1963, a windstorm took 22,000 lives. In 1965, Bangladesh was besieged with three separate storms which claimed another 57,000 lives. In 1970, a cyclone killed an estimated 300,000 people. Throughout the 1980s, storms and floods pounded the country, leaving death and destruction in their wake. By 2050, it is likely that 18 percent of the country will be under water. By 2100, some 38 million residents may have become refugees. The coastal mangrove forests, from which one-third of the Bangladesh national product is generated, will be lost.

A similar fate may be in store for large parts of Guyana, Venezuela, French Guiana, and the North Coast of Brazil. In Egypt, where population density is 700 people per square mile along the Nile river and delta, rising sea levels could submerge more than one quarter of the habitable land. Between eight and ten million Egyptians may become refugees.

One nation likely to disappear altogether is Maldives, a 115-square-mile chain of islands southwest of Sri Lanka in the Indian Ocean. The 202,000 Maldivians, living on 1,190 islands, have no high land toward which to retreat when the waters rise. Their entire nation is less than 6 feet above sea level.

All 169 islands of the Polynesian Kingdom of Tonga may disappear. Kiribati, Tuvalu, and Tokelau are nations and protectorates consisting of scores of small coral islands and volcanic atolls. Their total population is about 85,000 people. Nowhere are these islands more than 15 feet above sea level.

While refugees can retreat to volcanic slopes on Trinidad and Tobago, most of their fertile lands, which lie below the rising tide, will be washed away.

The 24 populated atolls in the Marshalls, like other islands in Micronesia and Polynesia, depend on freshwater "lenses" in the island interiors. Rising seas and storm surges will threaten these vital sources of drinking water. When the fresh water is gone, the people, too, will have to go.

Present laws prevent emigration from most of these former colonial nations but island residents who take matters into their own hands and become "boat people" will crowd onto other nearby islands which, while not submerged, will be economically stressed by increasing population, loss of production area, and storm damage.

AN OCEAN NEVER FORGETS

Ocean levels have always fluctuated in response to changing temperature. During the past ice ages, sea level dropped as low as 325 feet (100 meters) below where it is today. One thousand centuries ago, when the average temperature was about 2°F (1°C) warmer than today, sea level was about 20 feet higher. The reason that a two degree increase over the next few decades will not raise the oceans 20 feet is that the ocean takes much longer to heat than does the atmosphere. The circulation of the deep ocean currents is especially poor near the South Pole, which may serve to delay the melting of polar ice.

As the heat content of the atmosphere is pushed into the sea, the sea is slowly reducing its capacity to cool the air passing over it. The ocean will ameliorate the greenhouse effect by delaying its worst consequences, but this postponing process will also hamper efforts to moderate the greenhouse in the future. As the ocean absorbs heat and carbon dioxide, it reduces its own ability to absorb more heat and carbon dioxide.

If the water no longer becomes dense enough to sink as far or as fast as it does today, then the rate of removal of both heat and atmospheric carbon will decrease, and both heat and carbon in the atmosphere will increase even faster than projected.

If what we are seeing now in the ocean currents is a symptom of the greenhouse effect, as the evidence suggests, the worst-case scenarios are fraught with extreme danger. Ocean memory is both a comfort and a horror. It comforts us by slowing the effects of change, moderating our summer climate and giving us time to adjust. But the ocean is also like a six-diesel-engine train. Once it gets up to speed, it won't be easy to stop. When we speak, in hushed tones, of the possibility of a runaway greenhouse, we should recognize that the engine driving that runaway is ocean memory. The possibility cannot be ruled out that we are catastrophically flipping the ocean into an entirely different physical state. It is possible, maybe even likely. We just don't know.

Chapter Five
SUMMER HEAT

and absorbs heat much more effectively when sunlight strikes it than does the ocean. Land also gives up its heat much more quickly when the sunlight wanes. But land has memory, too.

Unlike air temperatures, which change hourly, temperatures beneath the ground surface maintain a running mean of the temperature swings at the surface. Because temperatures at the surface take time to propagate downward into the rock and soil, the deeper we measure temperature, the farther back in time is the interval of surface temperature history that it records. Deep soils filter out the high-frequency "noise" of short temperature swings between day and night, even between seasons of the year. The underground remembers the major events in surface temperature. In some cases, land's memory extends back for centuries.

Over the past few decades, scientists at the the U.S. Geological Survey have been measuring underground temperatures at various locations to get a better idea of surface temperature trends. They have concluded that careful measurements at depths of about 500 feet (150 meters) make it possible to read the temperature history of nearly anywhere on Earth for as far back as 2 or 3 centuries. To determine whether rapid greenhouse warming is really underway, they have focused their research along the Alaskan Arctic Coast from Cape Thompson to Prudhoe Bay, and inland from Barrow to the Brooks Range. Early findings show dramatic evidence of warming over the entire test area of 3,900 square miles (10,000 sq. km.). In northernmost Alaska, the temperature mean at the top of the permafrost, from 6 inches to 7 feet (0.2 to 2.0 m.) below the tundra, has increased 3.6°F (2°C) since the 1950s. The test data indicate that the warming started nearly a century ago in the Prudhoe Bay area of the Arctic Coast and began to climb into the Brooks Range only in the past decade. What the study proves is what we already know from other sources: a substantial change in the heat balance of the planet is now underway and it will be felt most dramatically in areas closest to the poles.

Most of the global climate models now being developed by the National Aeronautic and Space Administration, the National Center for Atmospheric Research, and others predict that the most severe greenhouse heating will be felt in the mid-continent regions of the Northern Hemisphere. There are at least two primary reasons for drawing this conclusion: studies of ancient climates and patterns of recent drought years.

The greatest uncertainty in the predictions is the change in ocean circulation that might occur as temperatures climb above anything we know from our historical records or studies of prehistory. If ocean currents shift, as some studies now suggest they may, so will the winds that moderate the temperature of the continents.

The extended periods of warm seas off the western coast of South America are often referred to as *El Niño* (the little man) or the southern oscillation. El Niño is used in Spanish-speaking countries to refer to the Christ-child. The reference to the warm Pacific originated because the phenomenon was observed to occur around Christmastime. Climatologists use the abbreviation "ENSO" to refer to **El Niño**-**S**outhern **O**scillation.

The North American droughts of 1982-83 and 1986-88 were ENSO-related occurrences. As sea surface temperature closest to the equator rose, the equatorial trade winds followed a particular pattern which carried them into North America in a steeper heading toward the north and east. The result was an "inverted U" shape as the jet stream either swung far to the north before encountering high pressure wind from the arctic, or split, with part of the jet stream swinging north through Canada and part following a southern route across the Gulf of Mexico.

As a product of this shift, the ridge of high pressure air that usually hangs off the east coast of Florida, the "Bermuda High," shifted to the southeast. The net effect, which we saw with regular frequency during the Midwestern droughts of the last decade, was a stable high pressure ridge in the center of North America. As the sun beat down, the soil heated. Hot air rose from the surface and augmented already high pressures at the upper altitudes. In this way, the pressure ridge tended to feed itself and remain in place, baking soils and shriveling crops. Without the Bermuda High to circulate moisture around the Gulf of Mexico and into the southeast, the Deep South and East Coast states from Arkansas to New Jersey joined the Midwest in experiencing a dry, unrelenting heat.

Every three to four years, El Niño spreads warm water across the Pacific and easterly winds slacken. Torrential rains and flooding batter the west coasts of North and South America. The Pacific Rim from Tasmania to Taiwan gets very dry. During the El Niño of 1982-83, loss of fishing and damage to crops cost Peru $2 billion and Ecuador $640 million. 600 people died in South America from torrential rains. During the same El Niño, Australia, Indonesia, and the Philippines suffered their worst droughts in history.

The reverse of El Niño, or a cooling of sea surface temperatures at the equator, has come to be called La Niña (the little woman). It was the La Niña effect of 1989 that preceded a late winter and an end to drought conditions in many parts of the Northern Hemisphere.

El Niño-Southern Oscillation

A warm current moving from west to east across the equatorial Pacific Ocean occurs at three to four year intervals. The movement of this current alters the normal circulation patterns of the prevailing winds, bringing drought to Southeast Asia, rain to the west coast of South America and warm weather to North America.

Source: Cromie, 1988.

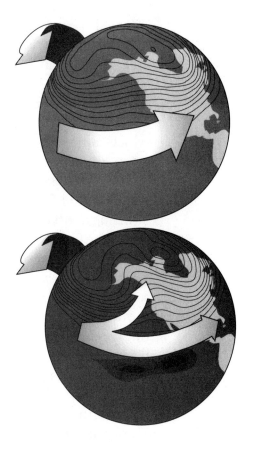

By measuring ENSO, oceanographers are now able to make predictions of general weather trends in North America some 3 to 9 months before the trends appear. El Niño's first appearance is marked by a subsurface heat wave, or ocean thermocline, that advances to the east at 135 nautical miles (250 km.) per day. It can easily be tracked by radar altimeters aboard weather satellites.

THE GREAT PLAINS

With significant global warming, it is possible that ocean currents will shift and the ENSO will vanish or will lose its predictive value. It might seem far-fetched to an Iowa corn farmer that the ocean currents off South America can affect his harvest, but the weather in much of the Midwest, as in midcontinent regions in other parts of the world, is directly tied to the ocean-atmospheric exchange. Weather arrives by wind, and the wind blows from the sea.

Sudden climate changes in local regions have occurred in the past and could occur again. The last major ice age which concluded around 10,700 years ago may have ended more abruptly in Greenland than elsewhere, possibly because of a sudden shift in the North Atlantic current. Recent examination of ice cores in Greenland indicates that sea ice retreated and Greenland warmed as much as 12.6°F (7°C) in the space of only 30 years. In those three decades, Greenland underwent more than half its total temperature difference between cold glacial and warm interglacial times. At the same time, dust concentrations in Greenland ice cores dropped to one-third of their earlier value, indicating a weakening of the storms that carried dust from North America.

The eleven states known collectively as the American West are, by nature, dry. In much of North America between the Mississippi river and the Rocky Mountains, the soil is thin and below it is only sand. As ice and snow gradually receded from North America 10,000 to 12,000 years ago, the changing climate was marked by frequent dust storms. That dust from the Great Plains can still be found in Greenland ice speaks mute tribute to the violence of the inland storms of long ago. And yet, as the Greenland ice reveals, eventually the winds grew quiet and the dust settled. Winter snows and summer showers moistened the prairie enough for the plants to return and hold the thin soil. Grasslands bloomed. Wildlife returned.

Major underground sources of water in the United States

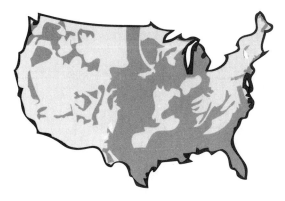

Source: *Geraghty and Miller, 1973.*

Prairie grasses are suitable for grazing herds of bison, but only if the bison keep moving and don't linger for very long in one place. By 1885, the Great Plains herds of wandering bison and burrowing prairie dogs had been replaced by more than 25 million cows and 50 million sheep. A century later, the livestock raised in the West reached one billion animals. Nearly 70 percent of the American West is now constantly grazed. Since they are not being hunted by predators, cattle and sheep tend to remain in one place until there is no more to eat. The earth they leave behind, bare of vegetation and compacted by hooves, is less able to retain water and reestablish its grass cover.

Late in the last century, the southern Great Plains developed into a dryland farming region dominated by the cultivation of wheat and corn (maize). To get the water for field crops and cattle, farmers drilled wells into the Ogallala Aquifer, a vast underground lake of water that extends from Texas to South Dakota and underlies most of Nebraska and parts of Oklahoma, Kansas, New Mexico, Colorado, and Wyoming. In this area today are nearly 100,000 farms covering more than 111 million acres (45 million ha.) and producing 80 percent of the milo, 40 percent of the wheat, 15 percent of the corn, and nearly half of the cattle in the United States.

Today most of the water used in irrigation in the Great Plains comes from groundwater, primarily the Ogallala Aquifer. The aquifer also supplies water to cities and towns, to rural homesteads, and to large industries. The largest user of Ogallala water is cattle. According to experiments performed at Michigan State University, it takes 2,500 gallons of water to produce an average one-pound steak. On a hot day, a cow will drink as much as 50 gallons of water. In order to

Projected losses of dryland corn production in the Eastern United States with a doubling of atmospheric carbon

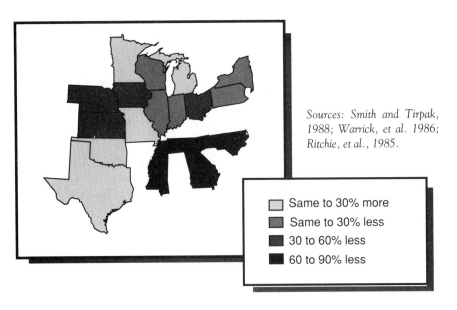

Sources: Smith and Tirpak, 1988; Warrick, et al. 1986; Ritchie, et al., 1985.

- Same to 30% more
- Same to 30% less
- 30 to 60% less
- 60 to 90% less

Projected losses of soybean production in the Eastern U.S. with a doubling of atmospheric carbon

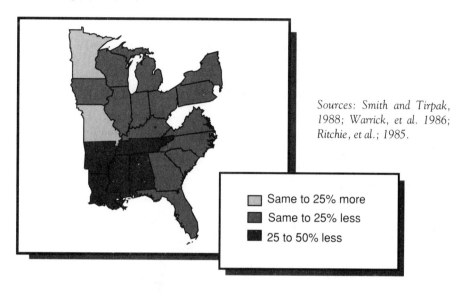

Sources: Smith and Tirpak, 1988; Warrick, et al. 1986; Ritchie, et al.; 1985.

☐ Same to 25% more
▓ Same to 25% less
■ 25 to 50% less

keep thirsty cattle happy, the Ogallala is being drained at the rate of 27 billion gallons per month, a rate that has caused the aquifer to drop up to 4 feet per year. Although water mining is a short term solution to the dry weather, it cannot be sustained for long. The Ogallala Aquifer is large, but not infinite.

From time to time, weather and soil erosion conspire to return dust storms to the North American plains. During the Dust Bowl of the 1930s, yields of corn and wheat were cut in half and more than a quarter million farmers migrated out of the region. During the drought of 1988, crop and cattle yields again plummeted, but many farmers remained, sustained in large measure by improved irrigation wells and billions of dollars in federal relief. As the Earth warms more rapidly in the coming years, the frequency of droughts in the Great Plains will once again increase. Both federal budgets and farmers will feel the strain. Overpumped aquifers will begin to dry out. The dust storms, perhaps even as bad as those recorded in the snows of Greenland more than 10,000 years ago, may return.

The two worst enemies of the farmer, after the weather, are insects and weeds. The increase in atmospheric carbon dioxide and the warmer temperatures will benefit both. Warming will alter the range and populations of pests, but insects tend to be very resilient and adapt well to changes which occur over the course of decades.

Changes in range of the potato leafhopper *(Empoasca fabae)* with a doubling of atmospheric carbon

Source: Smith and Tirpak, 1988.

The potato leafhopper is a major pest of soybeans. Present range is limited to Florida and the Gulf coast. With increased temperature and carbon dioxide, the potato leafhopper is expected to extend its range into several Southeastern states.

Field tests on plants grown under conditions of high carbon dioxide show that many plants grow larger and more quickly. Those that seem to do best include a number of common weeds, but also some valuable crops like wheat, corn, soybeans, and cotton. That's the good news. The bad news is that many of the plants which seem to thrive on increased carbon dioxide also suffer the worst damage from insects. The plant produced with more rapid growth is not as nutritious, either for insects or for humans. Insects have to eat much more of each leaf in order to acquire protein, which means more damage per insect and more damage per plant.

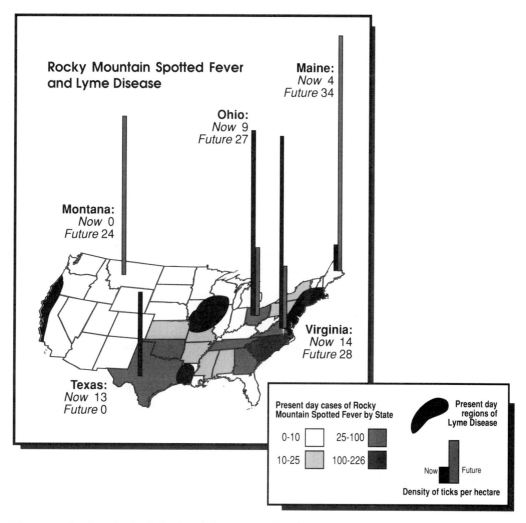

Changes in density in ticks found on hosts per hectare and states with highest current incidence of disease.

Changes in temperature and humidity will likely alter the ranges of bacteria and viruses as well as plants and insects. Both Rocky Mountain Spotted Fever and Lyme disease are already considered serious public health problems in the United States, but the spread of these tick-borne diseases is expected to rapidly accelerate with global warming. Rocky Mountain Spotted Fever is primarily borne by the American dog tick. With warmer winters and increased precipitation in the southern states, dog tick populations would shift from south to north. While this is good news for those in the south, the human population is much larger to the north.

Sources: *Smith and Tirpak, 1988; Wiseman and Longstreth, 1988; Haile, 1988.*

While the three principal climate models now used by the Environmental Protection Agency—the Goddard Institute for Space Studies (GISS) model developed by James Hansen and others, the Geophysical Fluid Dynamics Laboratory (GDFL) model developed by Wetherald and Manabe, and the Oregon State University (OSU) model—differ on prognosis for many regions, all agree that the outlook for the Great Plains is grim. While global warming will increase precipitation worldwide, the models suggest that the Great Plains will lose rain at a rate comparable to the drought years of 1934 and 1936—from 1.7 to 3.7 inches per year. Current models predict an average warming for the region of from 6°F (3.3°C) to 9°F (5°C), about twice as hot as it was during the Dust Bowl of the 1930s. That doesn't mean the thermometer will reach 200°F. It means that, on average, there will be twice as many days of 90°F or above.

Monsoons and Typhoons

While the Great Plains is likely to be one area where the impact of global warming will be most felt in North America, it is only one of many mid-continent grain and cattle producing regions of the world. Changing climate will also cause a northward shift in cultivated land, expansion of groundwater drawdown, and an increase in pest damage and economic stress in eastern Europe, the Ukraine, and across much of Asia.

In Southern Asia, the failure of the annual monsoon rains occurs only rarely, but when it does it spells catastrophe. Without rain, the food supply for hundreds of millions of people becomes threatened.

There is a competition between the warm Indian Ocean and the warmer land mass to the north. As the continent heats, hot air rises and draws in moist air from the ocean. The moist air then rises and forms clouds, which gradually build up to the monsoons. When the temperature of the Indian Ocean is slightly warmer, it creates a low-pressure area that draws clouds away from the continent. The monsoons never arrive. The threshold which the ocean must reach to steal the monsoons from the land is about 82°F (28°C). While this temperature has been reached infrequently in the past, it may occur much more often in the future.

A large body of water that may also go above 80°F for more of the year is the Gulf of Mexico. It is uncertain what effect that increase in temperature may have on the clockwise circulation pattern of ocean currents and clouds over the Gulf. We know that more frequent and larger storms are likely, but it is uncertain whether the storms will penetrate farther inland to damage—or refresh—the farming regions of Mexico, Latin America, and the southeastern United States.

Another open question is whether the warming of the Mediterranean Sea will translate into more frequent storms in southern Europe and northern Africa. It now seems to climatologists that the flow of air over the Mediterranean regulates the rainfall over both Europe and the Sahel, creating a mirror image in the area of 35 to 70 degrees North (mid-latitude Europe) and in the area of 5 to 35 degrees North (Northern Africa). As rainfall increases in one belt, it decreases in the other. Exactly why this occurs is as yet uncertain. What is certain is that the scope of the changes ahead is so large, the speed so rapid, and the degree of our knowledge so small, that very large changes in seasonal temperature and rain patterns may catch us quite unprepared.

Summer in the City

Cities are warmer than the surrounding countryside. Trees and grass evaporate water through their leaves, which cools the surface, but concrete, steel, and asphalt absorb and store heat. In the fall, frost warnings go out first to the suburbs and follow later in the inner city. In the summer, people often leave the cities to seek cooler breezes in the countryside.

This effect, known as the "urban heat island," has been observed by scientists for more than a century. It can be seen in towns as small as 10,000 people. The average heating is not great, only about 2°F (1°C) over the surrounding areas, but annual averages can be deceiving. With just a 5°F increase, a city like Washington, D.C., which now has one day each year of 100°F (38°C), would soon have 12 such days, and 87 days of temperatures over 90°F (32°C). Cities are likely to be one of the places where the extremes of the current global warming are felt first.

Days per year with greater than 90°F (32°C)

City	1950-1980	Doubled CO_2
Los Angeles	5	27
New York	15	48
Chicago	16	56
Denver	33	86
Washington D.C	36	87
Omaha	37	86
Memphis	65	145
Dallas	100	162

Sources: Hansen, 1987, Wilms, 1988.

Climate change will influence the demand for water in many cities. With hotter temperatures, people will water their lawns and house plants more, take more showers, and place more demand on public pools and baths. Climate change will also increase the demand for electricity, because people will use air conditioners and fans, and because refrigerators and freezers will have to work harder.

In North America, the heat wave of 1988 warped railroad tracks, which decreased train speeds, increased fuel requirements, and caused at least one major accident which injured 160 people. Steel expansion joints bubbled along interstate highways. The U.S. Army Corps of Engineers was forced to build a multimillion-dollar dike to protect New Orleans from saltwater intrusion up the Mississippi River—a dike that washed away when waters rose again in 1989. In New York, Nashville, and other cities, the heat of underground steam pipes combined with the heat of the sun to melt asphalt roadbeds.

Longer hot spells could cause trihalomethane formed during chlorination of city water to rise above allowable limits. Temperature rises may also increase hydrogen sulfide concentrations in metropolitan sewers.

INCREASED MORTALITY

About once every decade, a late summer weather pattern develops in the Los Angeles basin. A stationary high pressure cell over Nevada brings hot air, the "Santa Ana winds," to Los Angeles from the Southeast. The thermometer climbs over 100°F (32°C) and lingers there for several days. In late September, 1939, the recorded temperature at the Los Angeles Civic Center on 7 successive days was 100°, 103°, 104°, 107°, 106°, 103°, and 101°. In early September, 1955, the Civic Center thermometer read 101°, 110°, 108°, 103°, 101°, 102°, and 100°. Both of these week-long heat waves killed a large number of people. The 1939 hot spell produced 546 excess deaths, mostly among persons over age 50, before a tropical storm dumped 5 inches of rain on the city and brought an end to the sweltering heat. In 1955, the total excess mortality during the 8-day event was 946. The San Francisco earthquake and fire in 1906, by comparison, killed 542.

When the Santa Ana winds blew Nevada's heat into Los Angeles in late September, 1963, the population braced for another killer heat wave. The thermometer at the Civic Center climbed to 109° on the worst day, but the deaths which had been predicted did not ensue. Although there were 580 fatalities associated with the heat, the increase in Los Angeles population had led public health officials to predict 1,580 deaths. Looking at the records of the Los Angeles electric utility, epidemiologists discovered an all-time maximum in electricity use. Los Angeles had been saved by air conditioning.

Bans on CFCs in air conditioners, increased costs for electrical generation, and the rising price of food may create a hardship for many urban residents in the future. Given the choice between food and air conditioning, most people on fixed incomes will give up air

Air pollution for San Francisco East Bay with a 4°C increase in mean temperature

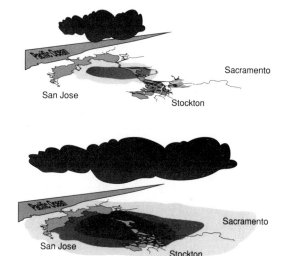

With increasing summer heat, the layer of city smog which builds up over Oakland, California and other East Bay cities on hot days will extend as far inland as Sacramento. Unless urban transportation is significantly restructured, peak ozone levels which now exceed maximum permissible concentrations (light shading), could exceed safe standards by more than 50 percent (darkest shading).

Sources: Smith and Tirpak, 1988; Morris et al., 1988.

conditioning. When the Santa Ana winds blow into Los Angeles in the 21st century, and when the mercury rises in other cities around the world, it is the poor and the elderly who are most likely to suffer and to die.

Planning for Change

As populations grow inside the world's major cities, as demands for water, food, and energy increase, and as the temperature rises, cities could become more violent, less hospitable, and less healthy places to live. It is possible to plan for the changes that are coming, but seldom do city governments plan very far ahead. Usually, it takes a crisis before the need for planning is even recognized.

It is possible for planners, elected officials, and city engineers to heed the early warning. But some people just have to learn the hard way. The speed of the changes that overtook Greenland some 10,000 years ago tell us that the changes now underway are not impossible. If we are reading them correctly, the signs tell us that the course we are now on will lead us to even greater changes, extending all over the world. Even if we were to alter our course now by changing the way we consume resources, many of those changes are locked in, will soon be upon us, and cannot be quickly reversed.

We need to begin to protect our sources of water, preserve our soils, and plan for a much more difficult future. What changes will occur in the large mid-continent regions of land as the greenhouse warms? We can't say for certain. It will depend to a great extent on what changes occur in the ocean currents. How will the ocean currents change in the coming decades? We don't know, but we have good reason to suspect they may change considerably.

Chapter Six

THE SKY IS FALLING!

In March of 1988, the U.S. National Aeronautics and Space Administration released a long awaited report of its findings on ozone depletion in the upper atmosphere. NASA had been sending a modified U-2 spy plane from Ames Research Center in California on trips up and down the Pacific coast to sample the atmosphere for two years. More than 100 scientists from 7 countries had spent 16 months carefully analyzing the data. The news was bad. Ozone depletion was not confined to the South Pole, but was found all over the Northern Hemisphere. Concentrations over North America were down 2 to 3 percent in the summer months and 3 to 6 percent in the winter.

The hole in the sky was first observed by a British Antarctic Survey team working at Halley Bay in 1982. The team, led by Joseph Farman, was using a spectrophotometer, a light-measuring instrument developed in the 1920s. The readings they were getting were so astonishing—a 20 percent drop in springtime ozone—that at first the team decided to distrust the instrument's calibrations until they could be verified. By October, 1984, after new equipment had been in use for a full year, they were sure. Ozone depletion over Halley Bay had reached 30 percent and was accelerating. Measurements from another station 1,000 miles to the north confirmed the finding.

Antarctica was the best place on Earth to measure changes in the ozone layer, because every year, between September and early November, a swirling mass of supercooled air—minus 110°F(-93°C)—called the circumpolar vortex, creates a virtual wall of wind around the region, isolating it from parts of the world which are being warmed by the sun and where fresh ozone is being created. In the Antarctic springtime, just before the first rays of the sun arrive, atmospheric ozone is at its lowest point and cannot be resupplied from outside air currents.

The discovery of a hole in the ozone layer over Antarctica came as a surprise to scientists and to the general public. There had been speculation since at least 1974 that chlorofluorocarbons (CFCs), a group of man-made chemicals widely used in aerosol sprays, solvents, and refrigeration, would have a long-term effect on the ozone layer, but the speed and extent of the damage came as a real shock.

In the late 1950s, Arctic ozone declined by 12 percent as a result of atmospheric nuclear weapons tests, but by the late 1960s, the ozone had fully recovered. At the other end of the Earth, Antarctic ozone was only marginally affected by the 477 atomic tests, which were forced underground by the signing of the Partial Nuclear Test Ban Treaty in 1963. Between 1957 and the middle of the 1970s, the concentration of ozone after the dark Antarctic winter (during which very little new ozone formed) was always about 300 Dobson units.

Normally, the polar vortex breaks down in November. When that happens, the Antarctic's ozone supply is replenished and the hole closes. In 1987, the vortex did not break down in November. It continued to swirl until mid-December. That delay reduced the amount of time for ozone resupply and also affected stratospheric wind patterns. By October 1987, the ozone measurements had reached the lowest level ever recorded over Antarctica, with more than 67 percent of the ozone destroyed in the spring. The spectrophotometer's gauge in Farman's lab at Halley Bay stood at 125 Dobson units.

Probing the Causes

When Farman's findings of a hole in the sky were reported in the May 16, 1985 edition of the British journal *Nature*, scientists at NASA's Goddard Space Flight Center began going over archived computer data to see if the discovery could be validated. Satellite remote sensing data, which had gone unnoticed because of the sheer volume of information being gathered, confirmed the worst. Since the 1960s, robot satellite sensors had tracked the formation of a hole in the ozone shield that by 1984 was larger than the United States and taller than Mount Everest. The computers should have been programmed to flash warning lights and sound sirens. Instead, the findings had been missed because the computers that processed the satellite receptions had been programmed not to report any measurement lower than 180 Dobson units, and to assume it was an instrument or radio-transmission error.

After Farman's report was confirmed by NASA, scientific work on the ozone depletion problem began in earnest. Teams of researchers arrived in the Antarctic to measure the hole. NASA began high-altitude flights to sample the troposphere. The news they began feeding back was grim. By the Antarctic spring of 1987, the average ozone shield over the South Pole had dropped to less than half its normal concentration. In some places it had nearly completely disappeared. The hole had also grown to more than double the size of the United States.

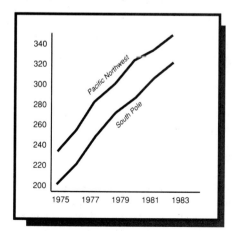

Increase in CFC-12 from 1975-1983 South Pole and Pacific Northwest in parts per trillion by volume

Source: Patrusky, 1989.

Moreover, the evidence seemed to suggest that indeed, as F. Sherwood Rowland and Mario Molina had predicted in 1974, the problem was caused by man-made chlorine compounds which had been accumulating for only about 20 years. Released into the air from bubbles in plastic foam, leaking refrigerators, abandoned air conditioners, aerosols like hairspray, spray paint, and insect repellent, fire extinguishers, and discarded chemicals, the gaseous detritus of better living through chemistry was threatening the existence of life.

CHLORINE

The sun emits a broad spectrum of radiation, only a part of which is visible as light. Radiant energy from the sun is carried to the Earth in particle-like energy units called photons. Photons of shorter wavelength (higher frequency) carry more energy, and fall at the ultraviolet end of the light spectrum. Photons of longer wavelength (lower frequency) have less energy and fall at the infrared end of the spectrum. If photons from the sun strike the Earth's atmosphere with enough energy, they split the molecules they encounter. That process causes the normal formation and destruction of ozone in the upper atmosphere. Ozone absorbs higher energy ultraviolet radiation and gives off heat. The amount of ozone in any area depends on the amount of radiation arriving from space (which depends upon the 22-year solar cycle and the rotation and orbit of the Earth),

Creating ozone from sunlight

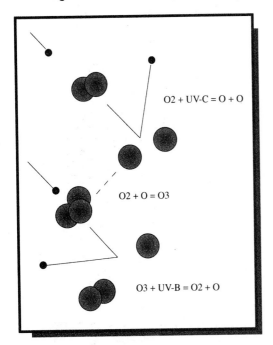

$O2 + UV-C = O + O$

$O2 + O = O3$

$O3 + UV-B = O2 + O$

In normal atmospheric chemistry, high energy ultraviolet photons break apart oxygen molecules, but the freed atoms of oxygen recombine to form ozone. Lower energy ultraviolet photons break the ozone molecules, yielding new oxygen molecules and free atoms, which can recombine to produce more ozone. Through this natural process of ozone formation and destruction, most of the harmful, high-energy radiation from the Sun is screened out before it reaches the Earth's surface.

the winds that transport ozone from the tropical latitudes to the poles, and, increasingly, the chemistry of the air.

Primarily as a result of human activity, chlorine atoms are finding their way into a region of the atmosphere, 6 to 30 miles up, where ozone does most of its work to protect the planet. One atom of chlorine can destroy thousands of molecules of ozone before being neutralized. Chlorine is neutralized by reaction with methane to form hydrochloric acid and carbon, which are eventually deposited back to earth by rainfall.

Chlorine monoxide is formed when an atom of chlorine destroys a molecule of ozone. The researchers at the South Pole discovered that concentrations of chlorine monoxide inside the ozone hole were hundreds of times normal. NASA flights to the North Pole reported similar concentrations.

The atmosphere contains very little chlorine produced by natural sources. Chlorine from man-made products—such as bleach and industrial compounds—usually combines with hydrogen in the lower atmosphere to form hydrochloric acid, which is washed from the air by rain. To reach the higher altitudes, chlorine atoms must be transported by a mobile carrier molecule. That man-made carrier is the one first identified by Rowland: chlorofluorocarbons or "CFCs." CFCs are everywhere. We in the developed world walk on urethane soles, ride on foam car seats, and sleep on pillows and mattresses of CFC-blown foam. CFCs are in the plastic peanuts that come out of shipping packages for TVs, VCRs, fine china, and other fragile goods. CFCs help sterilize surgical instruments, freeze seafood, and clean computer circuit boards. They are in boat-warning horns, car air conditioners, drain cleaners, spray confetti, photo-negative dust-offs and oven degreasers. CFCs are in the fluid coolant of nearly every refrigerator, freezer and air conditioner.

According to Mario Molina at the University of California's Jet Propulsion Laboratory, the generation and release of CFCs impaired the method by which chlorine is removed from the upper atmosphere. Molina believes that CFC collisions with ozone and other molecules produce an excess of nitric acid which serves to dissolve ice clouds. Without high altitude ice clouds to capture chlorine, each chlorine molecule may not just destroy thousands of ozone molecules, it may be free to destroy hundreds of thousands of ozone molecules. The loss of the ice clouds—the chlorine sink—when taken in combination with the commercial expansion of the CFC industry—the chlorine source—explains why the ozone hole grew so rapidly. "The Antarctic stratosphere crossed a source/sink threshold in the late seventies," says Molina. "It's been downhill since. And there's no reason to expect any recovery."

Destroying ozone with chlorine

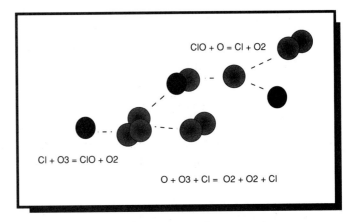

Chlorine depletes molecules of atmospheric ozone by turning them back into normal molecules of oxygen. One atom of chlorine is capable of performing the same process indefinitely. An average chlorine atom will repeat the process 100,000 times before being captured by a methane molecule to form hydrochloric acid and carbon, which are rained back to earth.

OZONE

Ozone, a three-atom molecule of oxygen, is one of the most important structural components of Spaceship Earth. It makes life possible. Close to the Earth's surface, ozone is familiar to many city dwellers as the principal component of "smog," the urban air pollution produced by exhaust fumes from vehicles and industry. The brown haze of oxides of nitrogen (which are reddish brown), ozone (which is blue-grey), and soot (of various colors) hovers over many metropolitan areas and can be seen from considerable distance, even from space. As smog, it is toxic and corrosive—something to be prevented whenever possible. However, ozone formed in the upper atmosphere, primarily by bombardment of oxygen molecules with radiation from the sun, is extremely valuable. The blue layer of stratospheric ozone screens out high energy ultraviolet radiation which would otherwise destroy life.

If all ozone above a particular spot on Earth were compressed to atmospheric pressure, it would be only a tenth of an inch thick. It is a very thin shield. Yet, the development of the ozone layer around the planet was a precondition to the development of multicelled plants and animals. If 20 percent of the world's ozone were suddenly to disappear, humanity would be eliminated, probably within a year. Two hours outdoors in direct sunlight would cause blistering of exposed skin. Food crops would shrivel and burn. Field mice and rabbits would be blinded, as would the hawks whose lidless eyes search the barren fields for their vanishing prey. The tiny shrimp and plankton at the surface of the ocean would disappear, and their loss would impair the entire oceanic food chain.

The phytoplankton at the ocean's surface may seem insignificant to most people, but over the course of a billion years, these creatures produced a significant portion, perhaps more than two-thirds, of the atmosphere's oxygen. Their role today is to absorb carbon dioxide and deposit carbon in the ocean, and to serve as the primary sunlight converter at the base of the marine food chain.

Inside plankton is a compound called dimethylsulfide, or "DMS." Phytoplankton probably synthesize DMS in order to protect themselves from the highly concentrated salt in seawater. The DMS escapes into the seawater when plankton die and is then passed to the atmosphere as a gas. Once in the air, it undergoes oxidation and creates an aerosol of sulfate particles that rises into the stratosphere. These particles are the grains around which droplets of rainwater form to wash carbon dioxide from the air. The more of these

Phytoplankton homeostasis

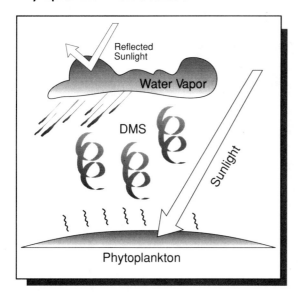

When sunlight warms the oceans, it stimulates the growth of algae and other microplants close to the water's surface. As these plants die, they give up their dimethylsulfide (DMS) to the atmosphere. The chemicals rise and form the sulfate nuclei of micro-raindrops. The more tiny DMS droplets in a cloud of a given size and saturation, the less sunlight reaches the surface. This cools the ocean surface and protects ocean plants from overheating.

DMS-origin sulfate particles that reach the atmosphere, the more sunlight will be reflected back into space and the more rain will fall. Recent studies of Antarctic ice cores confirm that production of DMS follows the 23,000-year cycles of axial tilt, indicating that DMS production by phytoplankton is aided or suppressed by the amount of sunlight reaching the ocean's surface. Phytoplankton reduce the warming effect of axial tilt by causing more sunlight to be reflected back into space. The phytoplankton, in addition to giving us oxygen and disposing of our carbon wastes, act as our planetary air conditioners.

Observations in the Antarctic in 1988 indicated that increased ultraviolet was killing colonies of phytoplankton close to the ocean's surface. Today the plankton are disappearing from the combined assault of ultraviolet radiation, acid rain, toxic waste, oil spills, and other human impacts. Yet, what we do to the plankton, we also do to ourselves. The loss of the phytoplankton at the surface of the sea is a bad omen for biological systems ashore.

Health Effects of Ultraviolet Radiation

On the average, a one percent depletion of ozone results in a 2 percent increase in harmful ultraviolet radiation (UV) reaching the surface. It now seems likely that depletion of the ozone shield could result in 5 to 20 percent more UV reaching populated areas within the next 40 years.

When ultraviolet radiation strikes a living thing, it is absorbed by the outer layers of cells. Microscopic plants and animals lack such protection. Many plants also react poorly to increased UV. Peas, beans, squash, cabbage, and soybeans show reduced nutrient content, slower growth, and lower yields.

In humans, exposed body parts generally have some protection from UV. Pigmentation in the skin stops UV-B from reaching lower layers. In the eye, the cornea and lens block UV from reaching the retina, the light-sensitive tissue that could be damaged by UV light. Chronic lifetime exposure to UV-B damages the eye's lens, contributing to the development of certain types of cataracts.

Exposure to UV-B radiation is linked to all forms of skin cancer, including basal-cell carcinoma, squamous-cell carcinoma, and malignant melanoma. The first two are quite common and usually curable. Melanoma, which represents only 3 percent of all skin cancers, accounts for more than 60 percent of skin-cancer deaths.

Ultraviolet Skies

UV-A: Ultraviolet radiation with wavelengths in the range of 320 to 400 nanometers. A low-energy radiation generally not associated with severe health effects, UV-A passes through the ozone shield and reaches the surface of the Earth with visible light.

UV-B: Ultraviolet radiation with wavelengths in the range of 290 to 320 nanometers. A higher energy radiation, UV-B is associated with deleterious health effects, and is only partially screened by the ozone shield.

UV-C: Ultraviolet radiation with wavelengths in the range of 40 to 290 nanometers. UV-C is a very high energy radiation and might cause severe health effects but is prevented from reaching the troposphere by the ozone shield.

With a 10 percent increase in UV-B, we could see a 7.5 percent increase in malignant melanoma and a 10 percent increase in basal and squamous cell carcinomas. Over the course of the next century, these skin cancers could afflict an additional 400 million people. Even with the best medical treatment, 7 million fatalities could result. An additional 80 million people could develop cataracts as a result of the change in outdoor light.

Another side effect of ultraviolet light exposure is suppression of normal resistance to disease—a loss of the immune response. In tropical countries, parasitic infections and infectious diseases are likely to increase as a result of the wet climate and increased ultraviolet radiation.

Phasing Out

CFCs are highly stable, virtually nontoxic, nonflammable, inexpensive, noncorrosive, and conduct heat poorly. They are excellent refrigerants and outstanding insulators. In the space of two decades, CFCs have entered a broad spectrum of commercial products. They are used all over the world. And yet, as word of the ozone hole spread around the globe, environmentalists, chemical manufacturers, and many government agencies began to come together and reach agreements about phasing out and eliminating CFCs. Potential replacements were quickly identified. Some chemical companies volunteered to scale back their production and use of CFCs and to increase research and development of alternatives.

Recently, more than 50 countries signed an international agreement—the Montreal protocol—aimed at cutting CFC releases in half by 1998. But in order to win signatures, the treaty included a number of loopholes which will permit nations to relax efforts in order to give more time for industrial adjustment to new products and technologies. While it looks good on paper and marks a milestone in global environmental cooperation, the Montreal agreement has to be seen as a beginning effort. The Montreal protocol will not stop ozone depletion, it will only slow its acceleration. In fact, the Montreal protocol commits the world to an increase in stratospheric ozone by the year 2020 of one order of magnitude—10 times the levels that existed before the treaty. Since CFCs persist in the stratosphere for more than a century, 90 percent of the molecules released in 1990 will still be doing damage in 2000, 39 percent in 2100, and 7 percent in 2300. *The damage we are doing to the ozone shield will be going on in the future for longer than the present ages of most nations.*

The replacements for CFCs are generally more expensive and less effective than the original compounds and so adjustments to alternatives will not be easy. Equipment using some of the replacements is also more expensive, will consume more power, and will wear out sooner. The United States, which currently produces over one million pounds of CFCs per year, about one third of the world supply, will spend $135 billion to replace its chemical production lines.

Americans will understand the real cost of the change when their 100 million home refrigerators, 90 million car air conditioners, and 100,000 office building air conditioners become unrepairable. While Vermont recently enacted a total ban on car air conditioners, that measure was met by skepticism in most other states and foreign countries. China and other countries in the early stages of industrialization have expressed unwillingness to forego the inexpensive refrigeration and air conditioning which the developed world has enjoyed for many decades.

Even if the Montreal protocol's terms are completely met, ozone loss will continue. Chlorine concentrations in the upper atmosphere will triple by the year 2075. Half of the increase will come from chemicals that are restricted under the treaty and the other half will come from chemicals that are not mentioned and from the liberal use of restricted chemicals by non-signatory nations.

In 1989, representatives of 124 governments met in London to discuss the seriousness of the situation. There was a high degree of understanding and consensus about ozone depletion. All participating nations agreed in principle that Montreal did not go far enough. Ozone-depleting chemicals must be phased out. The question left unanswered was how fast anyone was willing to act.

CFCs contribute to global warming in a number of ways. Like carbon dioxide, chlorine compounds serve to trap infrared heat radiating from the Earth's surface. Each CFC molecule is many thousands of times more effective at reflecting back heat than is a molecule of carbon dioxide.

By removing ozone in the upper atmosphere, CFCs allow more incoming sunlight to penetrate through the stratosphere and into the troposphere. Even if this additional light does not reach the surface, it serves to warm the atmosphere at lower altitudes, which has a global warming effect. As incoming ultraviolet radiation strikes oxygen molecules in the lower regions of the troposphere, it generates ozone there, and that low altitude ozone serves as an effective greenhouse gas which warms the surface. The effect of lowering the greenhouse ceiling may result in a much greater rate of warming than we have seen in the past.

The promise is that, given the frightening view from space of a gaping hole in the Earth's ozone layer, the nations of the world will find the resolve to come together and agree on an effective agenda for restoring our planet's ultraviolet shield. And if a complete ban on chlorofluorocarbons can be accomplished, perhaps we can reach the necessary agreement to limit and reduce other greenhouse gases as well.

Chapter Seven

TUMBLING DOWN

Some years ago Brazil recognized the value of the brazil nut tree to its national economy and passed a law prohibiting the cutting of the trees. As forests were cleared to make room for cattle farms, it became a common sight to see giant brazil nut trees standing alone in clear-cut pastures. Yet the law did not save the brazil nut trees. The trees in the clearings stopped producing nuts and became sterile.

The fertile stamens of brazil nut flowers are covered by a hood. Large Brazilian bees (*Euglossineae* and *Xylocopa*) land on the flower and enter the crack between the hood and its base. While they forage for nectar in the hood, their backs rub up against the fertile stamens and they become heavily dusted with pollen, which they carry from tree to tree. The flowers of the brazil nut tree need to be cross-pollinated by forest bees in order to bear fruit.

Brazil nut trees flower only in November. For the bees to survive, they need to gather nectar from a whole series of different trees, each flowering in turn, and each providing food for the bees in its season. The bees need a wholesome variety to get through an entire year. If any significant gaps develop in this variety, the bees leave and the brazil nut trees, as well as many other species, no longer produce.

The male bees are also dependent on orchids in the deep forest, which they visit at mating time. Rubbing against the orchids, they pack the scents on their hind legs and then fly off to form a lek, a group that attracts females for the mating ritual. If the orchids do not find suitable conditions for growth, they vanish from the forest, and with them go the bees, and with the bees, go the trees.

When brazil nuts fall to the forest floor their outer shells are eaten by large rodents, called agoutis (*Dasyprocta cristata*). The agoutis bury the seeds but often forget some of the caches. The agoutis' poor memory has the effect of dispersing seeds to favorable locations for new growth. The survival of the brazil nut trees is as dependent on the agoutis at the forest floor as on the bees in the tree canopy. If conditions are not favorable for any member of the forest community, all parts are endangered.

There are many examples of interwoven ecosystems. There are plants that are protected by birds, birds that are protected by hornets, hornets that are protected by trees, trees that are protected by fungi, fungi that are protected by ants, ants that are protected by plants. The demise of one plant species may eventually lead to the loss of up to 30 animal species because of the complex interplay of consequences. The global ecosystem is a complex interplay of symbiotic species, and all that inhabit this macro-network have evolved into a condition of interdependence, whether they recognize it or not.

Of all life on the planet, 99 percent is "phytomass" —a trillion tons of living plants. The desert and tundra regions contain about 2 percent of that, although they occupy a full quarter of the Earth's land surface. Human croplands account for only about one half of one percent of all plants, less than deserts and tundra. The bulk of plant life on the Earth—950 billion tons—is concentrated into tropical and temperate forests. Of this amount, well over half (about one-third of the land biomass) is contained in the tropical rainforests, which cover about 8 percent of land surface. Forests support greater diversity of life, and produce more of it faster, than any other ecological zone. A 4-acre patch of tropical forest may contain over 200 different species of trees. One species of trees may support more than 400 different kinds of insects. In the Amazon rainforest alone there are 80,000 species of plants and 30 million species of animals and insects.

The deepest forests of the world are million-year-old libraries of evolutionary knowledge. In some large rainforests are butterflies with 8-inch wingspans, caterpillars that masquerade as snakes, and trees that make their own insect repellent. Agronomists look to these and other unique biological wonders for future improvements to our supplies of food and fuel. Botanists recognize their value as sources of new medicines to cure arthritis, cancer, and mental disease. Locked in the genetic library of a humble jungle vine may be the magical solution to some great mystery of chemistry or biology. A vast library is now open to our understanding, but so dense are the varieties of life in the great forests of the world that all the zoologists, biologists, and botanists alive today, working together, would find that even to accurately survey all the diverse species is nearly impossible.

The number of species of living things in the tropical rainforests is in the tens of millions, but fewer than half a million plants, animals, and insects have been identified. The rest—untold plants, fungi, microorganisms, fishes, birds, reptiles, insects, and mammals—are completely unknown.

Forests also play a vital role in the planetary recycling of carbon, nitrogen and oxygen. They help determine temperature, precipitation and climate. They produce much of the cloud-cover that shields the planet from the full heating power of the sun. They regulate the greenhouse effect by drawing carbon dioxide out of the air. Forests, including their soils, store several times the carbon that is held in the atmosphere. With year-round growth, tropical forests can produce as much as 36 tons of carbon-sequestering plant material per acre per year, about twice as much as temperate forests can produce. In contrast, a hybrid corn crop grown in fertile topsoil can produce only 6 to 8 tons of plant material per acre each year.

Deforestation

Before the dawn of large-scale agriculture, some 10,000 years ago, 24 million square miles were covered by forest. Today, forest covers only 16 million square miles and that green area is shrinking rapidly. Since 1860, forest clearing has released between 100 and 200 billion tons of carbon to the atmosphere, roughly comparable to the amount of carbon released from the burning of fossil fuels. This year, another 2 billion tons will be released by deforestation in the tropical rainforest, as much as half of that from Brazil alone.

**Fuelwood demand in tropical countries
in millions of people overcutting or experiencing acute shortage**

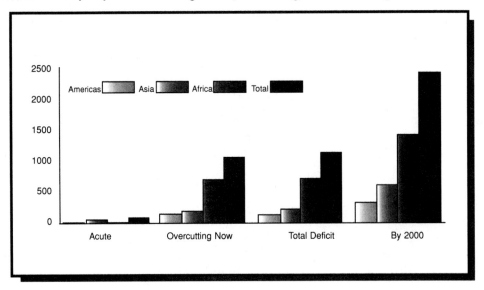

Source: Brown, et al., 1988.

Trees are vital to the survival of people who gather wood to cook and to warm themselves in cold seasons. But by the year 2000, the number of people lacking adequate supplies of fuelwood will reach 2.4 billion, more than half of the population in the developing world. Actual planting of fuelwood, mostly to meet the charcoal demands of urban areas, is only one-fifth of what it would take to replace what is now being consumed.

Most people are surprised to learn that the soils that support the most luxuriant tropical forests are often too infertile to support most forms of human agriculture. There are several reasons for this. Soils in temperate zones were remineralized during periodic ice ages, when massive glaciers scoured the landscape, digging up boulders, grinding them to dust, and fertilizing the soil. As the glaciers retreated, the land was reclaimed by plants which recycled the soil minerals and continued to build and rebuild the soil nutrients. Tropical forests are never glaciated. Occasional fertile regions exist near volcanos in Central America and Southeast Asia, but the volcanoes must be active enough to refertilize the soils periodically.

Most tropical soils do not have the benefit of active volcanoes. Once their soil minerals are leached away, they are gone forever. The life-giving minerals that are to be found in the tropical forests are in the plants and animals themselves. When leaves fall, or a tree crashes to the ground, it is decomposed immediately. It is completely gone in weeks. Its nutrients have been recycled into the forest, not left on the ground. The undisturbed tropical forest recycling program is extraordinarily complete. Almost nothing is lost or discarded. Rainforests flourish despite their soils, not because of them.

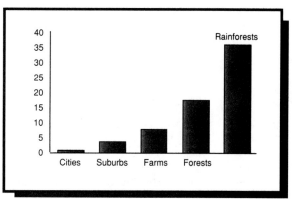

Comparative carbon absorption in tons per acre per year

Source: Myers, 1984.

In contrast to the richer temperate soils, many of the soils of the tropics are often difficult or impossible to farm. A garden in Brazil or Zambia yields about one tenth of the harvest of the same size plot in the fertile Ukraine or in the American Corn Belt. Many of the international developers who rushed to buy cheap farmland in Central and South America in the 1970s, and many of the government and banking officials who promoted massive rural resettlements of Third World farmers in the 1980s, were blissfully unaware of this simple truth.

In Brazil, 100,000 *seringueiros*—subsistence rubber tree tappers and nut gatherers—who had lived in comfortable symbiosis with their forests for a century, were suddenly displaced by internationally-financed cattle farms that burned the forests to grow grass for their expanding herds. Because of the poverty of the soil, each animal required 5 acres of pasture, and because the grass did not regrow easily, each animal needed a new 5 acres each succeeding year. As the herds expanded into the Amazon rainforest, valuable mahogany, teak, rubber, and brazil nut trees were bulldozed and burned. The forest nutrients—those that were not hauled away by commercial loggers—went up in smoke.

Many of these cattle ventures failed when their bank loans were used up. Ashes provided enough nutrients to farm for a year or two, but because most farming practices regularly break the soil and allow wind and rain to take it away, the scarce nutrients were soon exhausted.

In many areas, the land is too poor and expansion into the forest is too costly for any but the largest cattle ranching ventures—those that have their own bulldozers, meat-packing plants and ocean-going vessels. Whether or not rainforest cattle ventures are profitable is often irrelevant to the large landowners whose motivation is only to take advantage of tax-benefits provided for forest clearing. Frequently the forests are cleared without any attempt to farm the land. Yet the roads, gas stations, grocery stores, and communications brought in by the land developers favor the creation of new settlements by landless subsistence farmers, who come in to farm the land cleared and later abandoned, or to carve new farms out of forest that lies along major roads.

A survey conducted by the United Nations in 1982 estimated that tropical forests were vanishing at the rate of 70,000 square miles (180,000 sq. km.) per year, an area the size of Denmark every 12 weeks. Of that deforestation, commercial loggers claimed 18 percent, fuelwood gatherers 10 percent, and cattle ranchers about 8 percent. Most deforestation, about 63 percent of that occurring worldwide, was caused by subsistence farmers, some 300 million of them, who moved in behind the loggers and the ranchers.

The rates of deforestation have been growing faster every year. In 1987, 25,000 square miles was carved out of Brazilian rainforest. In 1988, it was 50,000 square miles. Indonesian forests, which are half gone now, will be completely logged by the year 2000. Virtually all lowland forests of the Philippines, Malaysia and West Africa have been heavily logged. Little virgin forest remains in Central America. Extensive forests of South America will, if current patterns persist, become cattle ranches and subsistence settlements. Half may be gone by the end of the 1990s. All equatorial rainforests may be gone by the middle of the 21st century.

Landless peasants understand that, despite the difficulties, their only hope of returning to traditional, self-sustaining, independent family and tribal economies lies in migration to "free" public lands in deeply forested areas. Consequently, the number of forest farmers can be expected to grow at a rate far greater than that of burgeoning Third World populations. Third World population is doubling every 40 years. The forest-farming population is doubling every decade. Because of the youthfulness of these populations, even family planning programs will not significantly slow this growth until well into the next century.

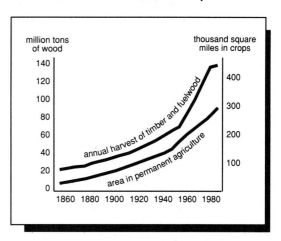

Deforestation in Southeast Asia in annual harvest of timber and fuelwood and forest land cleared for crops

Source: Houghton, 1986.

According to separate estimates by the United Nations, the International Institute for Environment, World Bank, World Resources Institute, Worldwatch Institute, and others, a century from now the population of coastal West Africa will have grown from 46 million people to approximately 260 million. Interior Brazil will have grown from 120 million to 280 million. The Philippines will have grown from 50 million to 125 million. Nigeria will have grown from 85 million to 530 million. Out in the rural areas, there may be few trees growing amid teeming masses of hungry people.

In the early 1980s, the government of Brazil, assisted by large international development grants and loans, constructed highway BR-364 from Vilhena to Porto Velho, in the far-western portion of the Amazon forest. Hundreds of thousands of landless families, encouraged by promises of free farming plots, migrated to settlements along the new road. The largest forest fire in history swept the hills and valleys of Rondônia, kindled by settlers anxious to clear land, fanned by the government authorities, financed by the international development banks. The clearing made by the fire was so large, it would have been visible from the moon.

When the smoke cleared, one third of the Rondônian jungle was gone, an area larger than Rhode Island. What was left was soil too poor to farm, streams choked with ash, and a fine yellow dust that covered everything—the soil of the rainforest. When the rains came, the dust turned to muck and the settlers, unable to grow their own food, were unable to transport any food in from the outside. Their trucks became hopelessly mired in mud. Brazil asked for and received a $200 million loan from the InterAmerican Bank to pave BR-364. With the road paved, the migration to Rondônia began in earnest and continues today, at the rate of 13,000 people each month.

In 1988, the deforestation of Western Amazonia was some 50,000 square miles (129,500 sq. km.), an area larger than either Honduras or Guatemala. Nearly half of this area was deforested to extend pastures for beef cattle, a consequence which some experts attribute to the increased demand for hamburgers for fast-food chains in the United States.

Today the governments of Brazil and other equatorial countries have become more concerned about the effects of deforestation. Brazil's supply of fresh water for eastern population centers is heavily dependent upon the Amazon water cycle. As trees are cut, the rate of rainfall is diminishing and river levels are dropping. Brazil is also one of the largest debtor nations in the world. Offers by international lending institutions to exchange "debt for nature," guarantees of rainforest preservation in exchange for a write-off of international loans, are now being welcomed by Brazil and other developing countries.

The debt-for-nature programs are small and experimental. They are not halting the logging, cattle-ranching, and subsistence farming incursions in any substantial way. Still, they hold promise if they could be expanded through greater participation by the International Monetary Fund, the World Bank, and others. Likewise, a program to plant a million trees in Guatemala, undertaken by a U.S. electric utility and several international agencies, while a useful experiment in the short-term, is only a band-aid approach to the larger problem. Certainly we need to plant more trees. But we need to plant more trees than we are cutting.

Earth's vanishing forest cover

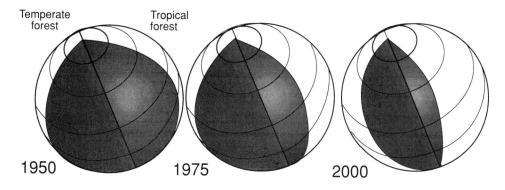

By mid-century, about a quarter of Earth's land mass was still forested. Today, only 20 percent is in forest. By the year 2050, experts predict less than 15 percent of the world's forest lands may remain.

Source: Myers, 1984.

On a global scale, the devastation of the forest cover has begun to disrupt the Earth's equilibrium. The loss of such a large portion of the Earth's trees means that less sunlight will be absorbed and more stored carbon will be released. Land that is fragile but currently maintained as forest through the secret synergies of countless cooperating species will be lost to life. For a short time it may provide grass for cows or sheep. Then, as we can see from our experience in Rondônia, the grass will stop growing. The land will dry up. It will become barren dust.

The destruction of the tropical rainforests has been likened to a giant walking across the land, flattening trees with each step—an area the size of a football field every second, the size of New York State every year, the size of France every five years. Each year, deserts claim another 15 million acres—23,000 square miles—beyond any hope of reclamation. Each year, an area five times the size of Switzerland becomes too impoverished to farm or graze.

EXTINCTION

How do the Earth's plants and animals respond to the loss of forest habitat? Biological communities are destroyed, ranges shift, and many species go extinct. Globally, biological diversity is diminishing.

Only five of the great extinction periods observed in geological history compare with the rate of extinction now underway. It has been suggested that the greatest extinction events of the past—the mass extinction when the world's landmasses were joined into the supercontinent of Pangea and the mass extinction at the Cretaceous-Pleistocene boundary (including the final disappearance of the dinosaurs)—were directly correlated to rapid increases in atmospheric carbon. Without any fossil air from those periods to examine, we can't know for sure. We do know that the new world record for global species extinctions in a single lifetime, or century, or millennia, will soon belong to us, those now alive, and this distinction will be achieved before the year 2000.

Biological diversity 600 million years ago to present in number of families.

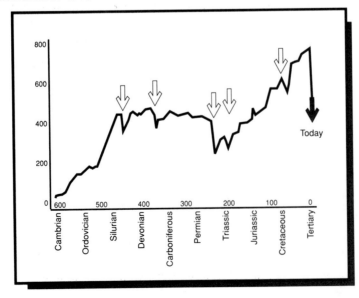

Arrows indicate mass-extinction events at the close of the Ordovician, Devonian, Permian, Triassic, and Cretaceous periods (declines of 12, 14, 52, 12, and 11 percent). In total numbers, the present decline (last dark arrow) is already unprecedented in Earth's history.

Source: Wilson, 1989.

Breakup of Ecological Communities

Because migratory rates differ, those species that migrate most slowly (dark shading) are left behind by species that migrate more quickly (lighter areas). For a time, there is a middle ground where species can still have access to each other, but as migration continues at a rapid pace, the middle ground is lost and the community dissolves.

Source: Lovejoy, 1989.

When the climate began to warm after the last ice age, biological communities were pulled apart but reformed into new configurations. Cold-climate species moved northward and warm-climate species expanded to fill the opening ranges. Time permitted a gradual evolution of new configurations of trees, shrubs, animals and insects.

This time, because of the speed of the change and the domination of the landscape by humans and their industries, species that can migrate poleward will find their route blocked by cities, roads, farms, and people. Biological communities will not have time to migrate in an orderly fashion and will be broken up.

THE RACE OF THE ROOTED ONES

In North America, there will be a tremendous displacement of forests as carbon dioxide doubles and temperature and precipitation change. Yellow birch, sugar maple, hemlock, spruce, beech and other temperate forests will need to migrate northward from 300 to 600 miles in the coming century. Many species of trees will go extinct throughout the Southeastern United States, but remnants of the species will still appear in the Great Lakes region and in Canada. Whether many species will reach safer northern climates is an open question. The fastest runner in the tree kingdom is spruce, which migrated an average of 120 miles per century about 9,000 years ago. For the slow moving beech, the rate was about 12 miles per century. Survival of these trees will depend on their ability to travel at least 300 miles per century in the future. At that speed, even spruce might not make it to its new habitat without help.

Human intervention in reforestation processes may not be possible. Manufactured new communities of trees, plants, and animals run the risk that unappreciated nuances of ecological interrelationships will be ignored, spelling doom for the artificial community. For successful engineering of natural areas, in-migration and out-migration of species must proceed at a more or less natural pace, acquiring its own balance. Migration strategies must involve establishment of protected corridors with modest transplantation of trees and slow-moving plants. The problem with the corridor approach is that on the scale that change is coming, the corridors must be extremely wide. Those corridors are now inhabited by people who have legally well-established property rights.

It is not known whether the adaptations of a Maine or Georgia beech tree to length of day, precipitation, and summer temperature are variable responses to local conditions or genetic adaptations of the species. If the adaptation is genetic, the northern trees are as endangered as their southern cousins. They, too, will have to migrate to cooler, wetter, or otherwise more suitable areas to the north.

The increase in carbon dioxide in the atmosphere, while favorable to photosynthesis, presents another hazard to migrating forests. Plants which respond best to carbon dioxide will outcompete and eliminate species with less favorable responses. Those plants with the quickest response are the weedy shrubs and grasses which directly compete with young trees in unforested areas. New forests will not be established easily, and the world will become much weedier. Parasites and pathogens will also be favored because they reproduce quickly and often become more biologically active as temperature and humidity increase.

In the United States, some 680 endangered plants may go extinct by the year 2000, about one every week. These include the flowering Hawaiian shrub, *Hedoyotis parvula*, found growing at the base of a cliff

Changing range of common trees in North America

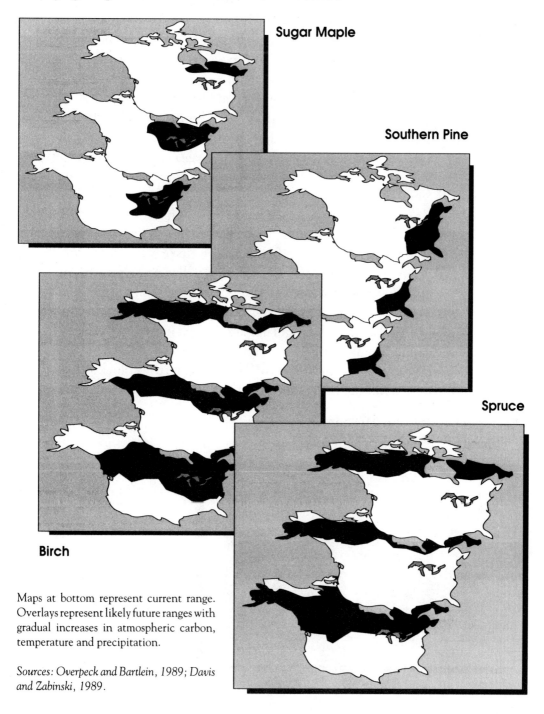

Maps at bottom represent current range. Overlays represent likely future ranges with gradual increases in atmospheric carbon, temperature and precipitation.

Sources: *Overpeck and Bartlein, 1989; Davis and Zabinski, 1989.*

on the island of Oahu; the brilliant yellow-orange *Amnisinckia grandiflora*, in California; the bright red cactus, *Opuntia spinosissima*, in the Florida keys; the Peter's Mountain mallow, *Iliamna corei*, in Virginia; the Texas flowering pods, *Lesquerella pallida*; and many other unique plants in habitats endangered by global warming, rising oceans, suburban development, and forest clearing.

The great sea turtles that cross the ocean to lay their eggs will come ashore on beaches that have grown warmer almost overnight. The temperature at which sea turtle eggs are incubated determines the sex ratio of the newborn. Warmer temperatures will mean fewer females, which may be enough to exterminate this species which is already at the margin of survival. Other migratory species depend on resources being at the right place in the right season, otherwise the migration breaks down. If warmer currents or ozone depletion deprive seagoing waterfowl of their food, they will vanish. If the herring don't rendezvous with the heron on schedule, both are imperiled.

The last time the world was 2 degrees warmer, tapirs and peccaries lived in Pennsylvania and manatees lived in the rivers of New Jersey. These animals are now trapped in tiny enclaves in the southern latitudes and cannot return. Their way north is blocked by civilization. They are as hemmed in as the southern forests.

Mississippi forests in the 21st century in tons per hectare of woody biomass

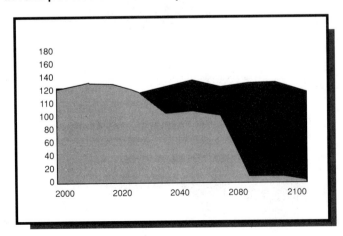

The dark shaded area represents the anticipated pattern of deforestation and reforestation in Mississippi forests without climate change. The lighter shaded area is the pattern of deforestation projected to follow a doubling of atmospheric carbon. Today Mississippi is 80 percent forested, but none of these forest areas are expected to withstand global warming.

Source: Smith and Tirpak, 1988.

The average duration of a species is some 5 million years, and for the last 200 million years, on average, 900,000 species have become extinct per million years, not quite one per year. The present man-made rate of species extinction is 10,000 times higher—higher than it has ever been before. According to the World Commission on Environment and Development, we are currently eliminating one species (plant or animal) per hour. By the year 2000, we will be eliminating one about every 15 minutes. Between now and then we could lose from 15 to 20 percent of the species that we know about. Up to now we have examined, in a small way, the physical features of about 1.4 million of the possibly 30 million species on the planet, and only 25,000 of the 250,000 species of higher plants.

Today most people derive their nutrition from no more than 20 species of plants, plants that were haphazardly discovered by our Neolithic ancestors and have been cultivated ever since. Botanists know of 75,000 more edible plants, some of them demonstrably superior to those we are accustomed to. For example, the New Guinea winged bean, *Psophocarpus tetragonolobus*, grows rapidly, reaching a height of 15 feet in a few weeks, and each part of the plant—roots, stems, leaves, flowers, and seeds—has a nutritional value equal to that of soybeans. So far, we have tested the edible qualities of less than one tenth of one percent of all the plants in the world.

Only in the past few years did we discover the energy potential of the babassú palm, *Orbingnya phalerata*, which can produce 125 barrels of petroleum-equivalent fuel per hectare per year. Only recently have we begun to cultivate the rosy periwinkle, *Catharanthus roseus*, which produces valuable medicines for the treatment of Hodgkin's disease and leukemia. That cultivation now yields $100 million worth of life-saving medicines each year.

Since 1960, 50,000 species have been destroyed in Western Ecuador to make way for banana plantations, oil wells, and human settlements. The medicines, foods, genetic immunities, and other potential benefits of these rare plants will never be known. In Madagascar, 6,000 plant species and 95,000 animal species have been exterminated in just 30 years. Five species of unstudied *Catharanthus* still exist on Madagascar, though some are now threatened with extinction.

All over the planet, this destruction is being repeated daily as the protoplasm in countless, diverse living things is consolidated into the human population.

Chapter Eight
Human Dimension

Human fallibility has always been a law of nature, but lately it seems to have been carried to an extreme. *Homo sapiens*, by virtue of its advanced cerebral electronics, opposable thumb, and supple larynx, has become a formidable parasite. It is recklessly wiping out all other lifeforms that are in competition for the natural resources that it eats, employs, or takes pleasure in. It is a reproductive wonder, doubling its population in the space of a single lifetime, while also extending its own life expectancy by medicine and, in the not-too-distant future, by genetic engineering. At the same time, it contaminates its own nest with foul and toxic wastes, it builds armies equipped with weapons that, if ever used, could annihilate itself. It contemplates its own destruction, and the destruction of life on Earth, as unfortunate risks that must be taken in pursuit of various abstract ends.

Have we always been this way, infused with a need to take risks much greater than ourselves? Have we an inherent need to dominate, to have dominion over all the Earth? Looking back to our origins, and on the long course of development that it took for us to arrive where we are now, it seems evident that we have always been blessed with advantages over any competing species. Do we compensate for those advantages by taking unnecessary risks to ourselves? Is our need for a sense of danger that overpowering?

EVOLUTION

The oldest known hominid remains come from Africa and represent a number of different species. Their genus is *Australopithecus* (meaning "southern ape"), rather than *Homo,* but they were distinctly non-apelike. They stood about 4 feet tall and their brains were about one-third the size of ours. One of the earliest of these hominids was *Australopithecus afarensis,* who walked fully upright by the good fortune of a double-curved spine. Most of what we know about *Australopithecus afarensis* we know from a single skeleton, about 4 million years old, found on the southern edge of the Red Sea in East Africa. We call her "Lucy."

Once Lucy stood upright, which is an uncomfortable posture for apes or any mammal with a straight spine, her forearms were freed for manipulating and inspecting objects. Lucy's orthopedic advancement led immediately to the evolution, in her descendants, of more nimble hands, longer thumbs, and sharper eyes. Those young hunters and gatherers with better hands, thumbs, and eyes had a survival edge over those without.

The more Lucy's hands gave her eyes to consider, the more information flooded her brain. Her descendants with larger brains were likely to be better able to assimilate more information and would tend to be favored with longer lives and more young who would inherit the larger brains.

Lucy's distant grandchild, *Homo erectus,* first made use of fire some 500,000 years ago. A female *Homo erectus,* with black hair and dark skin, living some 200,000 years ago in Africa, is now thought to be the common ancestor of all humans living today—a conclusion first suggested by biochemist Allan C. Wilson and co-workers after gathering representative tissue and hair samples from all the human populations of the world, extracting mitochondrial DNA, and constructing a line of maternal chromosome inheritance for the entire human family. We call our common ancestor "Eve."

It took 2 billion years from the birth of cyanobacteria capable of photosynthesis to the development of single-celled algae. It took another billion years before multicellular life sprang forth and covered the Earth. Higher brain function required several hundred million years to come into being. It took nearly 4 million years for humans to progress from the ability to walk upright to the ability to write. Lucy and Eve's distant grandchildren—those who could inscribe their religious visions on the walls of caves—made their appearance only about 50,000 years ago.

Around the time of Eve, or some time not long after, humans left their Eden of hunting and gathering and began to farm. One of Eve's sisters undoubtedly discovered that grains gathered for roasting and then spilled produced an abundant growth of those plants the following year. Eve's cousins learned to tame animals and keep them under control, even managing their breeding. The development of farming and herding made it possible to extract a much larger supply of food from a given area of land.

In *Homo sapiens sapiens* (the wiser wise man), Eve's heirs achieved the ability to speak, to learn more quickly and to record and extrapolate past experience into foresight. This qualitative endowment gave humans a quantitative edge in natural survival. They could transcend their immediate environment. It is this fourth dimensional brain, the one that recognized time and could articulate its reasoning, that now threatens all of the only known life in the universe.

After Lucy appeared, and for about the next 2 million years, the biological brains and bodies of her individual offspring approximately doubled in mass, while their group population remained relatively constant. A typical family group or clan, usually numbering from 12 to 24 hominids, might have had a range of no more than 15 miles, which they stayed within more or less constantly, until ice ages or other calamities forced them to migrate elsewhere.

Hunters and gatherers were free to move about, and had to in most regions, or they would have consumed all of the available food eventually. Those who herded were tied to their flocks. If they migrated, they had to migrate their flocks. Those who farmed couldn't move at all, or if they had to move, could only move back and forth by season. Once clans or tribes became stationary, it became necessary to protect their supply of food and water. Common defense required different musculature, newer tools, modified buildings and home sites, and greater social organization.

As one group grew in numbers and divided, a splinter group would occasionally emigrate into neighboring domains. If the splinter group had any genetic or cultural edge, it would tend to survive in the hostile new domain. If it did not, it would likely be extinguished. It was an either-or situation. Space and food technology did not permit many accommodations. Some tribes, and some species of hominids, remained hunters and gatherers. Undoubtedly, some hominids with less complex brain functions outcompeted other hominids with higher functions by sheer brute force.

Along with the development of writing between 40,000 and 50,000 years ago, humans gradually made another conceptual leap. An individual became able to maintain his social transactions as if he were still living in a small group, even though the group became much larger. For the first time in more than 3 million years, group size, land-occupying density, and human population as a whole, began to grow rapidly. Because these human tribes were reproductively efficient and had high survival skills, very early in human history all readily-available land became occupied.

Since that time, each doubling of population has been accomplished in half the time of the prior doubling, despite another ice age extending 10,000 to 20,000 years ago. We are now approaching the tenth doubling, a growth of one thousandfold, which is progressing a thousand times faster than it took 40,000 years ago.

Around the time the English landed at Jamestown, the Manchus invaded China, Tokugawa Ieyasu became Shogun of Japan, Cortez conquered Montezuma, and the Cape of Good Hope was taken by the Dutch, world population was about 500 million. Two hundred years later, it had not quite doubled, reaching only around 900 million. When many of those now alive were born a century later, the world population was around 3 billion. Today, not quite 50 years later, it is at 5 billion. In the lifetime of our children, it will double again, to around 10 billion. Barring some calamity like nuclear holocaust or a mutant virus, this is not a progression that might be, but a progression that will be. World population is growing now at the rate of about 1 billion every decade. As we start the 21st century, we will add 100 million new mouths every year. In the year 2000, the subcontinent of India alone will hold as many people as the entire world held at the outbreak of World War II. China will be even larger.

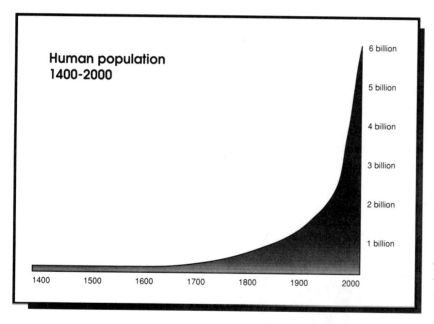

Source: Population Reference Bureau, 1989.

As societies become more developed and standards of living improve, the average onset of puberty comes at younger ages. This may be an adaptive genetic trait developed during periodic famines or between ice ages. This biological change helps to explain why there were 250,000 babies born to children in U.S. grade schools and high schools in 1989. Today girls born in North America, Europe, or Australia become biologically ready to procreate at the age of 8 to 12, a year or more earlier than their grandmothers.

Other species have high rates of reproduction, but nature conspires to keep them in check through a brutal process of elimination that begins even before they are hatched. Few if any of the other species inhabiting Earth have the survival and competition skills of *Homo sapiens sapiens*. As our medicine, science, and social welfare programs improve, as they inevitably must, the limits of population growth will have ever fewer biophysical impediments.

Still, people and cultures can and will die from overpopulation. They may not die from pestilence, epidemics of influenza, smallpox, diphtheria, bubonic plague or malaria the way they once did, but they will die from environmental contamination, contagious diseases, and famine.

THE NEED FOR FOOD

Two hundred years ago, Thomas Malthus predicted that at some point in time, human population will simply exceed the carrying capacity of the planet. It has reached, or will soon reach, that point. There are limits, and they are finite.

When the 1987 harvest began, world grain stocks totaled a record 459 million tons, enough to feed the world for 101 days. At the start of the 1989 harvest, carryover stocks had dropped to 60 days, lower than they were at the start of 1973, when short supply caused prices to double. We now have the shortest carryover supply since the period immediately following World War II. In 1988, the warmest year on record, North American grain production fell below consumption for the first time ever. Exports came entirely from reserves.

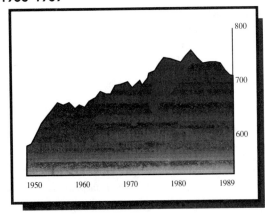

World grain harvested area in millions of hectares 1950-1989

Sources: USDA, *Worldwatch*, 1989.

Between 1950 and 1984, world grain production went from 624 million tons to 1,645 million tons, a 2.6-fold gain that increased per capita supplies by 40 percent. Since 1984, per capita production has declined every year, an average of 14 percent per year for each of the last 4 years. The reason for the decline is simple economics.

In the early 1970s, the Soviet Union and a number of other countries began to offset grain production shortfalls by buying grain in the world market. Grain prices doubled. Production surged. By the 1980s the world was awash in grain. Prices fell.

Some of the production gains came from overplowing and overpumping when prices were high, as well as from a frantic attempt to recoup investments as prices dropped in the early 1980s. Fence-to-fence farming depleted soil and water reserves, essentially borrowing from the future to produce as much as possible in the near term. As prices, and soils, declined further, farmers who could expand no more either cut back or went bankrupt. In the American Midwest, water tables are still falling 6 inches to 4 feet per year under a quarter of all irrigated cropland, meaning that still more cutbacks can be expected in the next decade.

World grain production in million metric tons 1950-1989

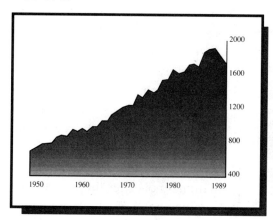

Sources: USDA, Worldwatch, 1989.

World grain production in kilograms per capita 1950-1989

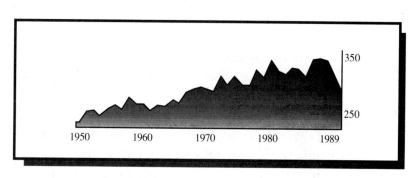

Sources: USDA, Worldwatch, 1989.

As food prices begin to rise in the 1990s, carried aloft on news of shortages, another boom in production is unlikely. The topsoils and irrigating water are becoming too depleted. Another green revolution of genetically engineered crops is also unlikely. If you look at U.S. corn yield per acre from 1950 to 1989, you see a gradual rise for 20 years and then a roller coaster effect that gets steeper, deeper, and more volatile as time goes on. One cause is the loss of genetic diversity in corn; the loss of resistance to variations in climate, pests, and other factors. Hybrid corn may produce more per acre in a good year under optimum conditions, but it can be totally wiped out in a poor year. The same can be said for a number of other "improved" crop strains, many of which are dependent on chemical fertilizers, herbicides, and other special conditions.

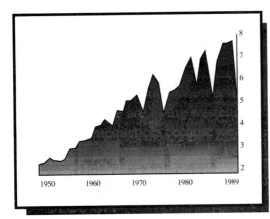

United States corn crop in metric tons per hectare 1950-1989

Sources: USDA, Worldwatch, 1989.

Because these miracle strains brought fast, short-term profits throughout the world, farmers planted them everywhere, replacing less productive native seeds that had, through centuries of evolution, developed broader immunities to pests, diseases, climate variation, and seasonal irregularities. This rapid loss of genetic diversity in food crops upon which we depend for our daily survival is more dangerous than many of us yet realize.

The optimistic assumption that humanity's salvation will lie in genetic engineering follows a recurring theme in agriculture and other human enterprises. We have a tendency to rely on the short-term technical fix without much concern for the long-term implications. Almost every human technical fix employed to modify naturally-evolved systems seems to reveal, usually fairly quickly, an inherent flaw that requires yet another human technical fix. Few human creations are as elegant as the developments of natural selection processes. By laboratory gene-splicing, we can reduce the time required for selective breeding of new plants or animals, we can create drought-resistant or heat-resistant strains, but we can't shortcut the viability/survivability time trials. Even if our new creations dominate the sprints, naturally-evolved species dominate the marathon events. They survive the hundred-year blight, the thousand-year blight, and the million-year blight. Few agronomists are yet willing to concede that to replace naturally-evolved genes with factory-made genes is to risk unwelcome surprises of substantial magnitude.

A second fundamental error implicit in gene-splicing is the presumption that breaking natural barriers is a great accomplishment. The genetic barrier that prohibits species from directly crossbreeding is a law that nature adopted early in the process of evolution. It meant that experimental lifeforms had to arise by the slower process of mutation and competition and that genetic diversity—which is a measure of health in biological populations—would be assured. Circumventing this ancient breeding barrier in the laboratory opens a door into a realm we have never previously entered. It is almost as foreign as if the laws of gravity and magnetism were suddenly suspended. For a brief moment in time, we inhabit a different world. But we should ask: Is it a better world? Is it sustainable?

**International crops
(U.S. Corn, U.K. Wheat, Japanese Rice)
in metric tons per hectare
1950-1989**

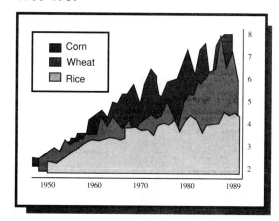

Sources: USDA, Worldwatch, 1988.

Before we move into the fast lane of genetic engineering, we might look at what the slower, but steadily progressing process of crop hybridization has wrought. The need for greater fertilization to take advantage of the new seeds we developed in the past half-century has directly affected the rate of global warming. Fertilizer production involves mining and processing of phosphate and nitrogen-bearing ores, and this mining and manufacture consumes fossil fuels and releases carbon dioxide, methane, and other greenhouse gases. Fertilizers also reduce the ability of soil microbes to remove carbon from the atmosphere. According to a 1988 experiment by the Woods Hole Marine Biological Laboratory, as much as one third less methane is removed by soils that have been fertilized.

Before the era of agricultural chemicals, farmers rotated crops to build up soil and control pests, or used a sustainable system of *permaculture*—relying primarily on plants that are hardy perennials.

Crops which are dependent on chemical fertilizers tend to rob the soil of its natural fertility, which increases the need for more fertilizers in succeeding years. As a result, after a number of years overall productivity declines, even as more fertilizer is applied. In North America in 1962, one ton of fertilizer yielded 10 tons of crops. In 1986, one ton of fertilizer yielded 5 tons of crops. In the 1990s, North America could double its use of fertilizer and there might be no measurable increase in crop yield.

Between 1985 and 1988, despite massive use of fertilizer, per capita grain production worldwide fell by 13 percent. Four-fifths of that loss was absorbed by reducing carryover stocks. The other fifth was absorbed by reduced consumption, particularly in the poorest parts of the world. Since 1980, per capita food production has declined 17 percent in Africa and 7 percent in Latin America, and these trends are accelerating.

Fully one third of the world's cropland is now in a negative soil cycle, each year retaining less soil productivity than the year before. Every year 24 billion tons of good farming soils—soils that took thousands of years to build—are eroded into the oceans. That is an area equal to all the wheat fields of Australia, each and every year. Expressed in productivity, a one-inch loss of soil equals a six percent loss in food production. Yet, growth of world population is bringing ever-greater stresses on marginal lands, with high rates of soil loss.

In China, as in the West, almost everyone wants their own family home. New homes are being built on dry, level ground, usually converted from farmland or forest. Almost all factories are built on former cropland. Since 1978, China has lost 7 percent of its cropland to homes and factories. In China, as in the West, irrigation is a good measure of food production capacity and the number of acres irrigated has begun to decline sharply in recent years. From 1950 to 1980, irrigated land expanded 56 percent in China. Since 1980, it has declined 11 percent.

World irrigated area in millions of hectares 1950-1989

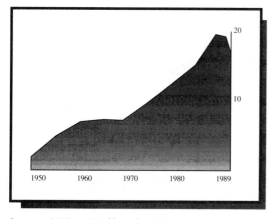

Sources: USDA, Worldwatch, 1989

As the climate across the mid-continent grain belts of North America and Asia becomes hotter and drier, wheat belts will move east. Forty bushel-per-acre wheat yields will replace 100 bushel per acre corn yields. Eventually, spring wheat crops of 30 bushels per acre may replace winter wheat crops of 40 bushels per acre.

In the past, the Midwestern corn crop has been two-thirds of the total United States harvest and one-eighth of the world grain harvest. Since 1980, three droughts have hit this enormous corn crop. In 1988, only 196 million tons were produced. During the same year, 206 million tons were consumed domestically. The 1988 export commitment of 100 million tons was filled from reserves. Corn prices on the world market increased by 50 percent.

OVER THE CREST

Global climate change poses a number of enormous problems for humanity an increase in skin cancer, dramatic sea-level rise, more violent storm damage, and droughts—all of which will result in a declining standard of living. Yet, these problems, while discomforting and life-threatening, are manageable in incremental stages; that is, manageable in the normal learning style of civilization. What is more serious, cannot be approached incrementally, and will be the first and most dramatic indicator of the deteriorating relationship between humanity and nature, is our supply of food.

We will soon find out, possibly within the next 10 years but certainly within the next 50, that we have exceeded Earth's human-support capacity.

In the 1970s and 1980s the world press often carried the image of a world awash in grain. Looking at rising production, surplus stockpiles, and depressed prices, economists were inclined to draw the conclusion, based on a typically short-range economic view, that production was plentiful and the only problem was one of distribution. This was a distinctly wrong conclusion. Reporters were missing the main story: a world borrowing from its children to feed itself, eating its seed grain, polluting its groundwater, and dumping irreplaceable topsoil into the ocean. Many ecologists saw something quite different: overplowing, overpumping, and overpopulating—trends that in the end can only spell disaster.

REAL FOOD FOR REAL PEOPLE

The original domestication of wild cattle in China and the Middle East was for ceremonial purposes. Over the course of several thousand years, cattle-keeping has been employed for food production in most cultures and has evolved into a ritual of its own. Even in Africa, Latin America, and elsewhere, the number of cattle a family owns is a mark of social status. In the North American middle class value system, steak is a symbol of success. As the proportion of animal products in wealthy diets has steadily increased, beef exports have become a vital source of income for lesser developed countries. As cattle production has increased, the producer countries, too, have developed an American taste for beef.

Livestock production is one of the world's most inefficient industries. Before World War II, most cattle were fattened entirely on grass. Today it is customary for steers to spend the last month or two of their lives in feedlots eating soybeans, sugar beet pulp, and molasses. It takes 6 to 8 pounds of grain to produce 1 pound of beef. On the way, the protein chains developed by the green plants lose up to 90 percent of their nutritionally useful amino acids.

Today, half of the world's grain—600 million tons of cereals—is fed to animals. While the United States devotes two-thirds of its cropland to produce animal feed and produces one-quarter of the world's red meat, the American appetite for meat has grown so dramatically in recent years that the U.S. is now a net importer. In

Latin America, Africa, and Southeast Asia, within the decade of the 1990s, the growing demand for meat may also surpass present levels of production for export.

The cow is a magnificent converter of otherwise indigestible cellulose, but the cellulose it is converting now is often tropical pasture that was only recently a rainforest. In the process of converting this cellulose, the cow produces methane, a potent greenhouse gas. The innocent-looking cow, standing in a field and chewing its cud, is at the center of the destructive ecological cycle that is precipitating global warming.

THE CRISIS POINT

We have now reached the point of diminishing returns for agricultural capacity. There are no vast undiscovered fields waiting to be plowed or turned into cow pasture. There are no miracle soil rejuvenators. There are no deep supplies of water waiting to be mined. Even if the next three or six or eight years are all good years for farming, it is unlikely that we will rebuild the surpluses of the early 1980s. Carryover stocks will probably hold to their present levels, or decline slightly.

When the next drought year comes, and it will as surely come as the sun will shine, whether it is this year, next year or in 10 years, the two-month supply we have now will ebb away and disappear.

What will it be like when this world food crisis arrives? We can make some educated guesses. The decline in living standards will be felt everywhere. At the lowest level of poverty, the price of food will rise above the power to purchase. There will be food riots in major cities. Feed for livestock will become food for the masses. At the upper levels of wealth, the pain will be felt, but it will be less acute. Other purchases will be curtailed to make up for a greater part of each household's budget being devoted to food.

At least in the near term, it is likely that household pets in America, Europe, and Japan will have a better diet than a starving child in Ethiopia or Bangladesh. Land that could be devoted to beans and potatoes will still be used to produce tobacco, alcohol, and flowers for the Tournament of Roses. Inevitably, such stark differences in purchasing power will deepen animosities and elevate world tension.

What is the carrying capacity of the Earth? Stanford University biologist Paul Ehrlich suggests that if the entire world lived at the food standard of the North America or Europe—40 percent animal sources of protein—some 2.5 billion people could be sustained by Earth's continuing productivity. If the world were to switch to the diet enjoyed in South America and many parts of Asia—about 5 percent animal sources—a population of 4 billion could be sustained. If everyone switched to a vegetarian diet—deriving proteins principally from legumes and whole grains—6 billion could be sustained. We will be approaching a population of 6 billion people by the year 2000.

Actually, a more sophisticated vegetarian regime has the potential of increasing carrying capacity above 6 billion. A key element in this approach is the adaptation of especially efficient protein foods such as soybeans, winged beans, and amaranth to traditional cuisines.

Soybeans yield 1 to 3 tons per hectare (0.4 to 1.2 tons per acre) in edible protein. If fed to cattle, it takes some 7 pounds of soy to produce 1 pound of meat. Grass-fed cattle under optimum conditions yield only 336 kilograms per hectare (300 pounds per acre). Soybeans can provide 10 times more consumable protein per acre of land than beef. Soybeans also fix nitrogen and are a valuable soil builder when used in rotation with other crops. Soybeans contain all 8 of the essential human amino acids, comparing favorably to meat, milk, or eggs. It is possible that if soybeans and other high protein plants were permitted to fill the gap, a human population of 10 billion could be sustained.

But the carrying capacity in terms of food supply is not the same as the overall carrying capacity. To sustain a life at a comfortable standard, more is required than an adequate supply of soybeans. There must also be clear rain, fresh air, wilderness, and peace. None of these may be possible to sustain with a world population of 10 billion or even 6 billion.

Human economies, since Babylonia and Mesopotamia, have been built upon the foundation of sustained growth. Economies have constantly expanded by one or two percent per year. To sustain that growth today, more and more national economies are spending their capital—stealing natural resources from future generations in order to buy groceries today.

The climate crisis is a crisis of human population. As the population of Africa acquires automobiles, it will use gasoline. For each gallon of gasoline consumed, 20 pounds of carbon dioxide will be produced. As the population of China acquires electricity, it will burn coal. Each kilowatt-hour of electricity generated will produce 2 pounds of carbon dioxide. As the population of Mexico consumes natural gas, it will produce carbon dioxide—12 pounds for every 100 cubic feet of gas. The American standard of living generates nearly 20 tons of carbon dioxide per person per year, and the populations of the world are demanding, and moving toward, the standards of living enjoyed by Americans.

As populations grow, they expand outward into farmlands, forests, and mountains. Wildlife vanishes. Rainforests fall.

Once human population hits the point of carrying capacity, only human potential, or quality of life, can continue to grow. However, at the point of carrying capacity, quality of life and population become inversely proportional. For the brain to continue to evolve, for life to become more pleasurable, for humans to continue cultural, social, and scientific advance, human numbers must decrease.

In the early part of this century the focus of learned discussions among futurists was not overpopulation or climate change but deterioration of heredity by the slower breeding of the genetically superior races. Between 1924 and 1932, a collection of essays titled *Today and Tomorrow* collected comments of notable British philosophers and scientists on the future of the world. Not one of nearly 100 contributors raised the prospect of overpopulation, although a number predicted widespread propagation of test-tube babies to make up for an anticipated decline in the natural birthrate. Many urged that these test-tube babies could improve the human situation if they were drawn from the right breeding stock.

Today, after the experience of World War II—where an attempt to genetically engineer the human race ended in holocaust—we recognize that breeding stock has nothing to do with our common dilemma—the differences between races, all of whom have common ancestry, being relatively minor and superficial—and that, on the contrary, concern with narrow heredity can be enormously destructive. Today we generally recognize that our greatest common problem is the sheer size of the human population. What is not well understood is how to cope with that problem.

We are now approaching the point where, in the next century, we will have to not just achieve zero population growth, but reverse our population trend and change from continually expanding to continually decreasing. The average number of young per couple must reach a unitary value, around 1.4. If we delay starting down this road for another century, or even for several more decades, the value diminishes still further, perhaps to 0.5, or one child for every two fertile women.

Any country that achieves replacement-level fertility is on the way to a stable population size. At replacement level, or zero population growth, births and deaths eventually balance out. With few exceptions, total fertility rates in the developed world are now at or below replacement level. England is expected to stabilize at 59 million people. France, Denmark, the United States, Italy, and West Germany may reach stabilization with another 10 or 20 percent of growth. These countries do not contain the bulk of the world population, however. What happens to them is instructive but may be inconsequential.

Five countries—Brazil, China, India, Indonesia, and Mexico—will account for nearly 40 percent of world population growth in the next 20 years. These five countries will add another 700 million people, about the population of India now. Among these countries, only China, with 2.4 children per woman, has a realistic prospect of achieving 1.4 births per family. In India, the average woman bears 4.3 children; in Mexico, 4.0; in Brazil, 3.5; in Indonesia, 3.3. While these figures indicate remarkable progress since 1960, when rates of growth were nearly twice as high, they are still too high for our planet's ecology to sustain. Moreover, for every success story in these nations, there are pockets of poverty where women bear more than 6 children on average. In Kenya, Afghanistan, Jordan, Tanzania, and Zambia, rates are higher than 7 children per family. In 15 Islamic nations, rates are between 6 and 8 children per family, a population doubling every 30 years.

Today the world must accommodate the equivalent of a new population of the United States and Canada every three years. If North American standards of consumption were universally achieved early in the 21st century, world carbon dioxide production could reach 110 billion tons per year, 20 times present levels. The Earth would experience a climate shock of profound proportions, beyond anything we are able to estimate at the present time.

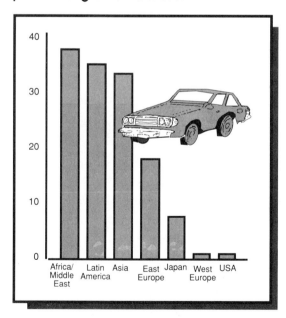

New automobile registration trends 1987-1992 in percent of growth from 1987

Source: *Worldwatch*, 1988.

While there may be more births per capita in Bangladesh than in the United States, the United States imposes more than 100 times the stress on the world's resources and environment for each child born. Babies in Bangladesh do not drive air-conditioned automobiles to the nearest hamburger stand, munch on grain-fed Brazilian beef with catsup and lettuce from Mexico, and drive home again. Central American forests are cleared to make inexpensive pet food and floral displays for North America. Papuan and Philippine forests are cleared to make inexpensive cardboard packaging for the Japanese VCRs in almost every American home. The rainforests are not felled to benefit the people of the tropical countries as much as to benefit the people of the industrial world.

What does this suggest about the levels of destruction we may expect when the developing world reaches a more advanced stage of consumerism?

Suppose China, by a remarkable program of providing incentives for small families and disincentives for large families, stops its growth at 1.1 billion (not 1.6 billion as projected). Suppose also that, by means of increased energy efficiency, China's population levels off its energy consumption at only 14 percent of the per capita use in the United States (versus 7 percent today). If the United States, the world's single largest producer of carbon dioxide, were to halt all coal burning today, amounting to 23 percent of its energy supply, it would not fully compensate for that small growth in energy consumed by each individual in China.

Suppose India, by measures no less innovative than China's, was to bring its fertility rate to 2.3 percent. And suppose also that India was able, by even more ingenious means, to level its per capita energy consumption at 7 percent of the per capita use in the United States (versus 3 percent today). If the United States was to halt all coal burning today, it would not fully compensate for that small growth in energy consumed by each individual in India.

Of course, these estimates are unrealistic. China and India will demand much higher levels of energy production and consumption than just a few percent of what the United States enjoys. Even if the average Chinese citizen were aware of the greenhouse crisis, which he is not, he is not at liberty to install photovoltaic cells on his roof or superinsulate his home. Even if the construction materials to perform that kind of conversion were available for purchase, which they are not, how could a person who makes $300 per year buy them? The fuel most affordable to the average Chinese is low-quality, high-sulfur coal.

As Paul Ehrlich says, "While overpopulation in poor nations tends to keep them poverty-stricken, overpopulation in rich nations tends to undermine the life-support capacity of the entire planet." Just as the developing world cannot possibly attain the extravagant style of wealth and wastefulness to which the industrial countries have become accustomed, so the industrial countries cannot, for either moral or climatological reasons, sustain that style either. The growth of human population is the most destructive facet of the problems now confronting us, but if those in the affluent countries can bring themselves to confront the enormity of that problem, there is hope.

The abundant diversity which existed a few short centuries ago has now become concentrated into one powerful monospecies, *Homo sapiens*, with a handful of domesticated animal and plant species kept around to serve that dominant species. We have cut down the carbon inhaling forests, paved over the carbon inhaling fields, and supersaturated the carbon absorbing oceans. We are creating more and more carbon exhaling humans, cars, factories, refineries, and power plants. We have cut into our ozone shield, carving out a vast hole. The Earth's hydrocycle has become our garbage dump.

Our biosphere, that part of the Earth which contains life, is as thin as the dew on an apple. Twenty miles below us, the Earth is white hot. Twenty miles above us, cold space.

Part Two

Chapter Nine
A Shift in Emphasis

or all our faults we are, we must admit, only human. We have our good traits. We are rational. We can change. If we can be brought to see what is needed, we can usually find a way to accomplish it.

The purpose of this book is to raise an alarm. We are too complacent. We should be frightened. This is a nightmare. We have to wake up.

We have options now, but our options will narrow quickly as time goes on.

Nothing is driving us over the cliff that is not of our own creation. Granted, we have heavily burdened ourselves with a panoply of counterproductive cultural metaprograms: acquire individual wealth;

have a large family; enjoy a thick sirloin steak; drive a sporty automobile; use electrical appliances—can openers, carving knives, toothbrushes—to replace human labor whenever possible; have several children, preferably sons; give the children pets to play with—dogs, cats, horses, parakeets. Many of these seemingly intrinsic human desires have only been with us a few decades. And we are virtually obsessed with a perceived need to instill these same aspirations in the next generation. Yet, we have the uniquely human capacity to change, to adapt to changes around us, and to make conscious decisions about what our lives will be like.

Some traditions—love of country, protection of property rights, pursuit of leisure time—may have helped us in our journey to this stage of understanding, but we have to begin to look back through them and to re-examine our assumptions at their roots. In the light of our impending demise, we have to find the holes in our mental, social, and scientific fabric that brought us to this peril and we must begin to repair them before it is too late. We must set about to consciously reassess all of humanity's behavior patterns with a new awareness.

If we continue as we are now going, we will soon be in the depths of a crisis which we may not be able to survive. If we modify our behavior incrementally—moderate our population growth, modestly increase energy prices, raise end-use efficiency, reduce carbon and CFC emissions, and undertake some reforestation—the result could be a substantial reduction in greenhouse gas buildup, but it would not prevent the coming climate crisis. With a concerted international effort of this kind, perhaps we could escape with a one-third increase in carbon dioxide by the year 2100, which is considerably less than the increase would be without the effort, but, because of the long duration of greenhouse gases in the atmosphere, reducing the rate of increase or even stabilizing the rate of emissions at current levels will not create a sustainable balance. To actually return to normal atmospheric equilibrium, a process that will take from 50 to 100 years even under the best circumstances, the world would have to reduce the output of carbon dioxide by 50 to 80 percent, the output of methane by at least 10 to 20 percent, the output of nitrous oxide by

80 to 85 percent, ban chlorofluorocarbons, and avoid any increase in emissions of carbon monoxides and nitrogen oxides. These goals are not unobtainable, but they require a much greater level of determination than we have demonstrated to date.

RESTRUCTURING

Until recently, human activities could be neatly compartmentalized within nations, within industries, within areas of academic study, or within other divided frameworks. That world is now gone.

The paradigm of manhood, harkening back to tribal traditions, is that of the rugged individual, stronger than adversity, fighting against hostile nature for his family or his clan. In order to better defend themselves, rugged individuals lent their muscle to the building of citadels for defense, to the creation of city-states, and nation-states, and multinational alliances, but only grudgingly. At each stage of the way, the lesser vehicle had to surrender just a little of its personal sovereignty in order to create the larger vehicle. Yet, deeply rooted in the common psyche remains the notion that one person, left to his own devices, should be able find the means to survive. That is the nobility-in-primitivism idea expressed in the writings of James Fenimore Cooper and Ernest Hemingway.

Alas, that paradigm is now behind us. To survive, it needed a wilderness larger than the human tribe. The tribe has dwarfed the wilderness. We have wrapped up the Earth in paper maps and divided it with deeds. We've paved Paradise.

It is not, as social commentator Bill McKibben says, the "end of nature." Nature will be back, and with a vengeance. It is instead the end of innocence. We cannot continue tampering with natural systems on ever-greater scales and expect to come away unscathed. Nature is poised to strike back, and what she will destroy first is the paradigm of the rugged individual. We will all be passengers in the same lifeboat, adrift on the same boiling sea.

Acquisition of wealth, which has been the perennial goal of most human institutions, does not provide an appropriate response to the challenges of the 21st century. Even if every husband and wife could afford to feed and house a large family, the planet cannot sustain that freedom. Even if every family could afford a personal automobile, Earth's fragile environment can no longer permit that goal to be accomplished. The balance between individual liberty and social responsibility is at the heart of the emerging debate.

The model of the solitary man as provider is shattered. The compartments through which we viewed the world before—nations, economic systems, academic disciplines, social or cultural similarities—are now dissolving. The environmental crisis, the development crisis, the energy crisis, and the adversarial relationships between East and West, North and South, are all part of the same vanishing paradigm. We cannot go it alone. We are all one. Our problems are the same. There is no way out for any unless there is a way out for all.

Getting it Together

It is easy to say that the uncertainties are so great at this stage that no action is warranted. We can ignore the early warning and just keep on going as we have gone on before. Perhaps people in Illinois would welcome a climate more like that of Georgia. Perhaps we can begin to grow corn in virgin soil above the Arctic Circle, or wheat in Greenland. Malaysian environmental minister James Wong has even found a good side to tropical deforestation: "If cutting rainforest trees causes warmer weather and less rain, it will benefit my golf game."

Some are inclined to more drastic technological fixes. They suggest we seed the atmosphere with carbon-absorbing chemicals, create high-altitude ozone with laser beams, ferry 35 million tons of reflective dust into the stratosphere, build orbiting "venetian blinds," or fertilize ocean plankton. We can expand our biotechnical capabilities to grow new foods in warmer climates, in poor soils, even on the surface of the ocean. Humanity is destined to be the master of its own environment, they argue, and although we are new at the art of weather modification, it is only a matter of time before we bring it under our control.

And yet, where the risks of mistake are so large, can we afford to expand our experimentation to an even greater scale? Are we confident that we know what we are doing to Earth's infinitely interconnected systems?

Jean Jacques Rousseau, in his *Discours sur l'origine de l'inégalité des hommes* (1754), supposed that "natural man" was neither good nor bad, but that by imposing restraints upon himself by dint of tradition and education, had given up his innocence and become his own worst oppressor. The way out, Rousseau argued, was for society to reform itself by establishing a universal social contract that balanced individual liberty against the collective needs of all humanity. The greatest collective need of humanity, Rousseau said, was to develop a more profound respect for nature and for natural systems.

The challenges now spread before the world are challenges to governments, to religious institutions, to economic frameworks that underpin international trade, and to individual people of widely varying traditions, cultures, aspirations, and views.

These challenges cannot be met by more of the same, or by the next best technical fix. Neither perennial optimism nor endless expansionism will do. These challenges require a fundamental restructuring of the human relationship to the natural world.

Chapter Ten

Deep Ecology

In the Third Century B.C., Aristotle proposed a unified system of human relations that never really caught on. According to Aristotle, all human beings have the same fundamental needs: food, clothing, shelter, love, intellectual challenge, and so on. We all have similar desires, urges that lead us toward happiness and fulfillment. Knowledge, skill, and pleasures of the mind are unlimited goods—those things of which you cannot have too much. Other goods are limited. If you have more wealth and bodily pleasure than you need, Aristotle cautioned, it may not be good for you, or for the rest of us. Because of varying needs and desires, and the distinctions between them, tensions between people inevitably arise. It was to deal with these tensions, to create a fair resolution or system of justice, that we formed societies.

Aristotle said that the underlying purpose of any society was to provide justice. Justice requires each person to help all others in obtaining the real goods needed for a good life: enough food, shelter, and other basic necessities to be healthy and vigorous. The Aristotelian concept of natural rights is that we have a right to those things we really need and it is the duty of society to provide us the fair means to acquire them. To Aristotle, a just society is a society that provides the basic requirements equally to everyone, and also makes available equal opportunities for individuals to pursue their own happiness in the development of greater knowledge, skill, or other human pleasures, according to their individual talents and inclinations.

When natural rights are impinged, as we have seen in every civilization, people suffer inequitably. Creativity and natural productivity are stifled, turmoil grows, and eventually the multiplying dissatisfactions erupt into violent acts of crime, rebellion, and war. When natural rights are liberated, as they have been for small groups of privileged people in different eras, there is a blossoming of creativity, productivity, and enlightened thought. The modern political world seems to be locked in a constant struggle between advancing natural rights and betraying them for the sake of more immediate, pragmatic goals that require exploitation of the weak by the strong.

There is a cultural bias, found in all parts of the world and extending back even before Aristotle, which regards humans as the crown of creation. In recent years, this anthropocentric (human-centered) tradition has been challenged by a more radical, biocentric (life-centered) movement known as deep ecology. The term "deep ecology" was coined by Norwegian philosopher Arne Naess in 1973, but it took on new, and uniquely American, connotations in the environmental debates of the 1980s.

Today's deep ecologists—who find their roots in Aldo Leopold, Henry David Thoreau, John Muir, Aldous Huxley, M.K. Gandhi, George Santayana, and others—assign to the human species a smaller, more humble role in the natural order, generally adhering to the ethical prescription of Leopold: "A thing is right when it tends to preserve integrity, stability, and beauty of the biotic community. It is wrong when it tends otherwise."

It is a tenet of deep ecology that untouched wilderness is of the highest possible value and therefore requires the highest order of protection. A related principle is that all human inventions are merely discoveries of the ways of nature and that by study of nature, humans can find harmonious existence and contentment. In the words of Lao Tsu, "Man follows the Earth, Earth follows Heaven, Heaven follows the Tao, Tao follows what is natural."

The non-Aristotelian concept of natural rights includes the rights of those with four legs, and fins, and wings, and roots in the ground. It includes the rights of senior species to habitat and survival. It encompasses all our relations.

In the 1980s, diverging environmental values created considerable controversy in the political arena of North America. Thrusting deep ecology into the spotlight was a protest faction that advocated direct action to save virgin stands of temperate rainforest timber in the Pacific Northwest that were threatened by corporate raiders who bought large timber companies in leveraged buy-outs and clear-cut the forests to pay for the takeover. "Monkeywrenchers" were no longer content to block bulldozers at the edge of old growth forest. They drove five-inch spikes into trees and covered the nailheads with pine tar. When the trees reached the sawmills, the saws were damaged or destroyed. Monkeywrenchers also spiked roads and drilling equipment, pulled up survey stakes and traplines, downed billboards and powerlines, and otherwise obstructed the march of civilization into wild areas.

In contrast, more traditional environmentalists, who rose to new levels of official recognition and influence in the 1970s and 1980s, advocated protection of natural resources for basically utilitarian reasons: preservation of scenery, the "wilderness experience," and biodiversity for the sake of present-day research and recreation and for the less easily defined benefit of future generations.

This traditional anthropocentric viewpoint is under attack from within the environmental community because it reduces natural areas to finite resource reservoirs which can be drawn down as needed to meet defined objectives, and those objectives may bear little relation to the habitat requirements of other species.

Many deep ecologists are more inclined to argue that humans are a disease in the living body of the Earth. The role of deep ecologists should be that of antibodies, doing what they can to preserve natural systems while providing time for Earth's own resistance to build and to eventually cure the disorder, removing humans if necessary.

But the threat to our global commons—our favorable climate for life—is not reducible to a debate between anthropocentrists and biocentrists, a debate which only a small part of the world's teeming population would even consider relevant. The ecology of the Earth is threatened by pollution from the industrialized world and from urban elites in the Third World, by growing militarization and nuclearization, and, at a more fundamental level, by three underlying diseases: economic inequality, political oppression of dissent, and overconsumptive, up-the-stimulus lifestyles.

To the poor, the oppressed, and the marginalized, the deep ecology ethic borders on the absurd. In parts of India and Africa, poaching tigers, elephants and other endangered animals inside of game preserves is considered as high a calling as is monkeywrenching in the American Pacific Northwest. Those displaced by designated wildlands and national parks view nature reserves as belonging mostly to a feudal elite, to foreign banks, or to rich tourists.

And yet, deep ecology raises a challenge that cannot be ignored. It asks the harder questions: Is evolution headed in a particular direction? Are we headed that way now? Can we, or should we, change the trajectory?

Looking back over all of human history, we should be groping with the difficult question of where we went wrong. We must have programmed ourselves incorrectly at some point, or we wouldn't be rushing headlong to our own extinction.

The living Earth seems to have a feedback mechanism that funnels individual organisms, species, and families toward planetary homeostasis: if they deviate too far they encounter environmental resistance. If they adapt successfully, they minimize resistance. This is the evolutionary impulse, the Tao of Nature.

Humans departed from their evolutionary path when they began to set up their own feedback loops to enable greater deviation from natural systems. They created the technical fix. If more fields were needed for agriculture, we cleared the forests. If stone tools were too limiting, we first forged iron, and then steel. If overuse of a particular pesticide fostered an immunity in the pest species, we made a stronger pesticide.

The creation of climate-altering technology was a response to the demands created by ever-greater deviation. Each technological fix we added seemed to create more problems to be solved. We continue to devote more people and more resources to think about these problems and to develop technological solutions. And so we grow, and become more capable. But what happens if we reach the limit of our resources? What happens when we pass the point of no repair?

When we begin to reach the edge of natural limits, the first thing we notice is the resistance of nature to the deviation. It was there all along, but as we lose the ability to maintain our overstretched artificiality the friction begins to assert itself once again.

Advanced technology may have the capacity to postpone environmental resistance, but it does not, and should never have, the capacity to eliminate it. In only a few years time, we may find that the majority of the world's people are reduced to a standard of living lower than that of a century ago. Is this possible? Yes. Is it likely? We don't know. The longer we are unable to sense the Earth's resistance, as we carry on with life as usual, insulated by our car air conditioners and cradle-to-grave insurance policies, the less we will recognize the full extent of our deviation, and the greater the stored-up disequilibrium will become. Eventually, even our strongest dikes will cave in under the pressure. The insurance premiums will cost more than anyone can afford. All of our advanced technology will not solve the essential problem.

In 1889, Hiram Maxum developed something he thought would end all wars. He invented the automatic machine gun. No one would be so foolish as to attack a country with a line of these around its borders, he said to André Maginot. World War I undid Maxum's maxim. Harry Truman's faith in the atomic bomb met a similar fate. Ronald Reagan was so certain that Star Wars technology would end the threat of nuclear holocaust that he was unwilling to trade it away, even in exchange for the elimination of all nuclear weapons.

Around 1780, when Ned Ludd led groups of English handworkers into the steam mills and smashed the stocking frames and power looms that were taking away jobs, the political issue was the legal or moral right of factory owners to employ mechanical inventions to replace traditional craftsmen. The demands of the Luddites—for democracy in the workplace, quality handwork, and protection of women and child laborers—became synonymous in the popular press of the day, and in the historical connotation of Luddism, with blind opposition to progress through industrialization.

Mohandas Gandhi was labeled a "Luddite" when he complained, a half-century ago, about the British power looms that were destroying the employment of masses of traditional weavers in Madras. Gandhi countered by arguing that it was as much a sin to overproduce as to overconsume, if by that overproduction you stole another's opportunity to work.

Hierarchies of wealth and power

In the pyramid model (left), a few people control vast wealth while most people are in a condition of extreme poverty. In the diamond model (middle), social engineering takes from the wealthy and gives to the poor, creating a large middle class. In the hourglass model (right), vacillating policies create a large number of very wealthy and very poor, and relatively few in the middle.

Market system democracies in general tend to evolve hierarchies of wealth and economic power on one end and poverty and servitude on the other. With regulation by government the "pyramid" of wealth—a few wealthy at the top tapering down through varying degrees of wealth to a large number of very poor at the bottom—becomes a "diamond" distribution—with few very rich and few very poor and a large middle class. In the more prosperous Western nations, which have vacillated between market intervention and free market, progressive taxation and subsidy, austerity and liberal spending, an "hourglass" distribution of wealth—large clusterings at the top and bottom, and relatively few people in-between—is evolving. Of these three models, pyramid, diamond and hourglass, the diamond is generally what is sought. Today, for the majority of the world's people, the international economy tends to resemble a pyramid.

Those at the bottom of the pyramid are becoming increasingly convinced that poverty is not an aberration but a premise of the current economic order. And yet, without the support of these people, the international effort to arrest global warming will be of no avail.

Marketplace adjustments, government initiatives, and common efforts to curb waste will all help slow the rate of climate change. These are modest and prudent steps that will buy us time. What we must do with that time is to think deeply about major shifts in economic structure, about changing our philosophical relationships to nature, and about what might be required for real sustainability as a species on this planet.

REDEFINING NATURAL RIGHTS

When Thomas Paine, Thomas Jefferson and others argued about natural rights, they were trying to establish the idea that human beings had an inherent predisposition to freedom of thought, expression, and action that was not granted to them by the King of England or by any governmental charter. Many modern democracies began as experiments in that vein. At this point, governments are no longer the problem—humans are. Other species have natural rights that humans are stepping on, although, of course, humans are stepping on their own rights in the process.

Sixty-five million years ago, 60 to 80 percent of the world's species disappeared in a cataclysmic mass extinction, possibly caused by an asteroid's impact with Earth. Human population, not an asteroid, will cut the remaining number of species in half again, in just the next few years. The natural rights of those species count for nothing in the macroeconomics of our times. Their natural rights are ignored completely.

The loss of biological diversity is tragic because it destroys totally the product of millions of years of painstaking evolution. It is tragic because the full value which we forever deny to ourselves cannot even be calculated. But beyond these aspects lurks an even greater tragedy. The loss of species is an early warning of our own vulnerability. We do not understand the web of life. Are we like the brazil nut trees, dependent on the forest bees, the seed gathering rodents, and the orchids? Are we now standing alone in the middle of a pasture, proud of having dominated the forces of nature and survived?

The time has come to break with past patterns. All nations have to secure peace, internally and externally, by leveling standards of living. That is the foundation of environmental protection. We have seen in the past that, most often, the culprits of environmental damage are the political and economic underdogs—the small homesteaders moving into Amazon rainforest, the poachers in African game preserves, those searching for firewood or places to graze herds in poor desert lands. The institutions and practices that create victims, renew poverty, and suppress human rights must change. The real world of interlocked economy and ecology will not.

There can be no solution to the climate crisis unless the world can come to fundamental agreement and can find the resolve to take decisive action. No single country or activity is the dominant source of greenhouse gases. No single remedial measure will have a significant effect on stabilizing the global climate. But in combination, broad worldwide strategies to reduce greenhouse gases, if they begin at once, have the potential to limit the total warming and to begin reversing the warming process.

To weld that commonality of interest we have to adopt the entire human family as our own and share our resources more equally. We have to listen to the advice of Aristotle who, more than 2,000 years ago, said that fulfillment for human society consists in giving freely to each person the bare necessities for health and well-being, and in providing each with the opportunity for self-improvement. This approach is perhaps too anthropocentric for many deep ecologists, but it is inevitable that, as we come to grips with the problem of human dimension, a guaranteed minimum standard of living will come to be seen as a necessary prerequisite for stemming the climate crisis and preserving the ecology of the Earth.

Chapter Eleven

The New Agenda

A child born to a family in an industrialized country will use 30 times more of the world's resources before he dies than a child born to a family in a lesser developed country. You cannot tell the people of the Third World that they may not have automobiles, electricity, or refrigeration when a disparity like that exists.

History contains many examples of civilizations that rose to great heights and then crashed into extinction by exceeding their resources. When they reached a certain size they began to spend their capital, as if the growth they had accomplished could be sustained indefinitely. Those civilization crashes should be important lessons to us, if we have a mind to think about them. No ancient armies, no matter how strong, were powerful enough to feed themselves when the forests were cut and the soils were depleted. No seafaring nations could trade enough to feed their people if, by overuse, they lost the resources from which their base of trade derived. No matter how great or powerful a sovereign might become, the basis for that greatness must come entirely from nature's sustained support.

The propensity to live on the earth's capital—its fossil fuels, forested land, freshwater aquifers, and abundant wildlife—rather than depending upon the regular income derived from the sun and from the Earth's regenerative systems creates deficits so severe in some sectors that the system's regenerative powers can be overwhelmed.

Today one billion people are living in poverty. Their numbers are growing. That poverty undermines the world's natural resource base. It cuts 150,000 square kilometers of forest, spreads 60,000 square kilometers of desert, and removes untold thousands of plant and animal species from the Earth each year.

**Child mortality by nation
in deaths younger than age 5 per 1,000 live births**

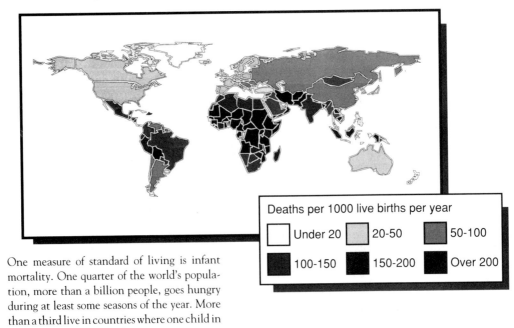

One measure of standard of living is infant mortality. One quarter of the world's population, more than a billion people, goes hungry during at least some seasons of the year. More than a third live in countries where one child in ten dies before the age of 5.

Source: *United Nations, 1988; Clark, 1989.*

Sometime in the next 40 or 50 years, the Earth's population will double, from 5 to 10 billion people. The world economy, in order to continue at it's present growth rate, must multiply 5 times. What will happen when the four-fifths of the world's population, many of them growing up on a visual diet of television shows like *Dallas* and *Wheel of Fortune*, ask, respectfully, to have the same standard of living as the other fifth?

Greenhouse warming as a function of population growth

$$\text{Total CO}_2 \text{ Emission} = \frac{CO_2}{\text{technology}} \times \frac{\text{technology}}{\text{capita}} \times \text{population}$$

This formula shows the buildup of carbon dioxide (and would apply equally well to other greenhouse gases) due to the size of the human population and the limits of technology. Although improvements in technology can limit carbon dioxide sources and improve carbon dioxide removal, those improvements depend on both the degree of distribution of the technology and on the size of the world population.

Source: Ehrlich and Holdren, 1989.

The remedy for our mounting crisis is simple. *We have to live within the means that the planetary life-support system provides.*
The countries of the world that can and must lead the way to a sustainable future cannot be hypocritical about the efforts of others. Today the United States and others are exerting great pressure on Brazil, Guatemala, the Philippines, and other equatorial countries to spare their tropical rainforests from destruction. At the same time, the United States is cutting its own tropical forests in Hawaii, Florida, and Puerto Rico at an unprecedented rate. In the past half century the U.S. harvested nearly one third of the old-growth timber in the Pacific Northwest; millions of acres of 1000-year-old stands of Douglas fir, Sitka spruce, hemlock, alder and cedar. The North American temperate rainforests, combining good soil with constant rainfall, still contain more biomass than all the tropical rainforests of the

Amazon. In recent years, the cutting of old growth timber in the Pacific Northwest has not abated, it has accelerated.

Carbon dioxide emissions by source nation

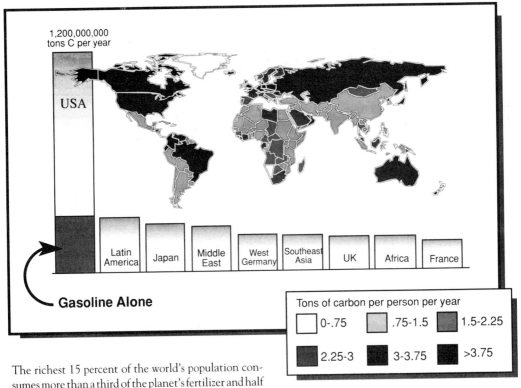

The richest 15 percent of the world's population consumes more than a third of the planet's fertilizer and half of the energy produced. Consequently, releases of carbon dioxide from energy use, industrial activities, and deforestation are largest in the major industrial countries. The United States produces more carbon dioxide from automobile fumes than is produced from combined sources in most other regions.

Sources: United Nations, 1988; Subak, 1989.

There is no manual that comes with Spaceship Earth. We have to discover its mechanics. We have to set aside the time to learn. It is an investment of time and effort we cannot afford to forego. It will require a careful program of instrumentation and constant monitoring from the depths of the ocean to the edge of space. It will require preservation of the data gathered for an extended length of time, probably centuries. It will require assembly of that information into useful and accessible forms. It will require analysis, thought, and reanalysis.

The men and women who gathered the data on levels of atmospheric carbon at Mauna Loa Observatory, beginning in the 1950s, were modern heroes. They had to conquer lack of funding, apathy, boredom and scientific disinterest. But they may just have saved the world. The men and women who trekked to Halley Bay to read the ozone measurements, month after month, year after year, were heroes too. These are recent discoveries. We had no idea of the peril we were in. What else should we know?

The second task we must undertake is to eliminate the waste and mismanagement of resources. That task will require reassessment of our styles of living, particularly the lifestyles now taken for granted in the industrial world. The United States contributes about 25 percent of each year's accumulation of atmospheric carbon. The Soviet Union and the eastern bloc together make up another 25 percent. Western Europe produces about 15 percent, China about 10 percent, Japan about 5 percent, and the rest of the world—some 180 other countries—contribute the remaining 20 percent.

While institutions usually change slowly, they can change quickly if the needs of the times demand. When a nation goes to war, everything changes, virtually overnight. Present necessity overwhelms inertia. The abandonment of feudalism in Japan and Russia, the end of slavery in Europe and America, and the international decision to stop most forms of whaling are all examples of sudden, large-scale changes that were brought about by changes in social conscience.

All of nature's systems are closed loops. Oxygen is exhaled by plants and inhaled by animals. Carbon dioxide is exhaled by animals and inhaled by plants. Heretofore, human economic systems have tended to be linear—resources are gathered and spent. When they're gone, we explore for more. That linear accounting system may have worked when the resource was deer or elk and the size of the human hunting tribe did not exceed the herd's ability to reproduce, but now we are spending our capital as if it were income—wiping out the breeding stock of the herd. In order to cure this imbalance, honest bookkeeping must add in the costs of sustainability—it must close the loop. *People must pay the full cost of replacing the resource before they consume it.*

No one likes governments to raise costs. If given the vote, people will tend to vote for smaller taxes and greater individual control over spending decisions. Even when a minority are deeply hurt by economic imbalances, the majority has generally tended to vote for policies that benefit themselves. In the context of global climate alteration, such artifices as "majority" and "minority" have to be seen for what they are. Those outside the national boundaries and in future generations do not vote, and yet they may be both the majority and the most adversely affected. Even if they cannot participate, they have to be made part of the process. They have to be protected.

Governments of the wealthy nations must lead the way toward sustainability in international economic systems. After World War II, the United States created the Marshall Plan to facilitate reconstruction of Europe. Today the United States spends nearly half as much—$45 billion in 1988—protecting shipping in the Persian Gulf. What is needed is a world-scale Marshall plan for creating sustainable economic systems. In the United States, the Environmental Protection Agency should have a budget larger than the Department of Defense. In the United Nations, the U.N. Environmental Program should have a budget larger than the Security Council.

When the United Nations Environmental Program was established in the early 1970s, it was given a broad mandate to stimulate, coordinate, and provide policy guidance throughout the United Nations. It was given a small office in Nairobi, two advisory boards, and a voluntary "target" budget of $100 million for 5 years. Since 1980, the UNEP has managed to raise about $30 million annually, primarily from the United States, Japan, the Soviet Union, Sweden, West Germany, and Great Britain. That amount that has not increased at all in recent years, while the range of tasks and activities assigned to UNEP have grown considerably. Thirty million dollars is about what the United States Department of Defense pays for two light helicopters.

The money currently budgeted by the United Nations for environmental programs is clearly an inadequate sum. To raise more money for this effort, the U.N. must have some new source of revenue, such as an international tax on global resource exploitation—a users' fee for the global commons. Nations could be assessed according to how much carbon dioxide they add to the atmosphere, how much toxic waste they dump into the ocean, how many acres of forest they clear, and how much fossil fuel they drill and mine. National governments can raise these international payments by taxing them at the source.

In March, 1989, leaders of 17 countries and high-ranking representatives of 7 others convened in The Hague to address the global warming crisis. For the first time, world leaders met and reached agreement that some degree of national sovereignty would have to be sacrificed if the planet were to be saved from environmental devastation. The Hague Declaration called for a new institution within the United Nations system that would have real power in coping with climate change. Unanimous votes within the proposed new environmental security council would not be required for action. Violations of its decisions would be referred to the World Court. Although the United States and Great Britain have thus far withheld support for a new international agency, the Hague Declaration was endorsed in principle in June, 1989, at the Economic Summit in Paris.

There is one obstacle, more than any other, that stands in the way of bringing human population under control and creating sustainable energy, agriculture, and other supply lines. That obstacle is debt. Most developing countries carry more debt to the wealthy nations than they can reasonably expect to repay. In 1988, the most indebted nations repaid $43 billion more than they received in aid. In essence, the world's poor were net exporters of money to the world's wealthy. This is not a sustainable economic order.

People do not consider the global environment when they do not have adequate food, clothing, shelter, and fuel for cooking and warmth. Floods and famines, plagues and wars do not significantly slow population growth. Raising individual standards of living does.

Raising the average standard of living while decreasing the rate of resource consumption and environmental pollution is the challenge of the 21st century. Whether we succeed in meeting this challenge may well determine whether we are still here in the 22nd century.

Chapter Twelve

TWENTY-ONE BETTER IDEAS

It is still possible, by acting soon enough, for both industrialized and developing countries to limit the buildup of greenhouse gases without abandoning other important social goals. The uncertainties about global warming—which are not, after all, uncertainties about whether the Earth is warming, but about how fast, how much, and where it will warm—do not justify delaying immediate action to improve education, reduce fossil fuel consumption, or embark upon reforestation. All those activities are completely justified, in their own right, for separate reasons.

In both the industrial world and in the developing world, population controls are justified for political and health reasons. Energy efficiency improvements are justified in order to maintain competitive strength in international trade. Efforts to halt soil loss and deforestation are justified for food and fuel considerations.

It is unrealistic to suggest that the impending climate crisis can be avoided without difficulty. To escape the worst of it, we will have to somehow bring together widely divergent cultural, nationalistic, and economic interests, to alter human behavior in fundamental ways, and to persuade a majority of the world's people to sacrifice their own potential comfort and prosperity for the sake of those in future generations. These obstacles are not insurmountable, but they are formidable.

POPULATION

Human population growth cannot be greatly curtailed without individual incentives to limit family size. Where there are high incidences of infant mortality, excessive rates of disease and death, oppression of women, inadequate education, pervasive poverty, and severe restrictions on individual self-determination, there are inevitably high rates of birth. The procreating process is genetically driven. It is organically pleasurable. Birth too, despite the physical effort, is a joyful experience. Children are a happy blessing. Where there may be few other pleasures in life, the poor will always have those. When a woman has no other source of fulfillment, she has that. It becomes a vicious cycle, because although children may bring pleasure close at hand, they create and renew the conditions for poverty.

To stem the rising tide of humanity, the world must acknowledge the disparities between the haves and the have-nots and must seek to redress that central imbalance. Small donations of money and technology are not enough. There must be a serious effort to level the minimum standards of living enjoyed by all.

1. PROVIDE UNRESTRICTED FREE ACCESS TO BIRTH CONTROL.

At the same time that we strengthen the social, cultural, and economic motivations for limiting family size, creating a demand for birth control, we must provide the means. Ninety percent of Third World women over age 21 either want to stop childbearing altogether or to delay the birth of another child. Each nation must give free access to sex education, contraceptives, and reproductive planning support services. The World Bank estimates that if $8 billion per year were provided to the poorer nations to enable them to create and dispense simple information services like leaflets, workshops, and classroom instruction, world population growth could be stopped at 10 billion, 2 to 6 billion lower than it might otherwise become.

The United States, the Soviet Union, and other developed countries could demonstrate leadership by adopting a two-child family as a national goal. Establishing that simple standard would encourage international institutions to adopt a similar goal for the world as a whole.

After World War II, faced with the constraints of its loss of empire, Japan cut its population growth rate from just under 2.2 percent to scarcely 1 percent. It did this by creating a national family planning program and providing contraceptives, information, and counseling.

In China, the government made a decision in the early 1970s that the national birthrate should be reduced. The policy was called *wan-xi-shoa*, meaning "later-longer-fewer." Although the state family planning agency coordinated efforts, it delegated the creative initiative to the provincial governments. Sichuan province began making monthly payments to couples who agreed to limit their family to one child. The 5 yuan per month that the province paid was approximately equal to a family's monthly earnings from agriculture, so families that joined the program doubled their standards of living. As the Sichuan

Birth and death rate in Japan in births/deaths per thousand in population 1948-1989

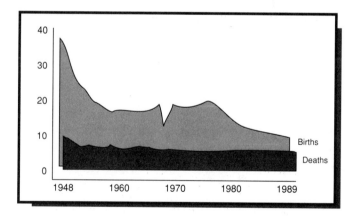

Sources: United Nations 1988, Worldwatch 1989.

program was replicated throughout the nation, China's population growth dropped dramatically. Today more than 73 percent of Chinese couples use modern contraception, 5 percent more than in the United States.

**Birth and Death Rate in China
in births/deaths per thousand in population
1948-1989**

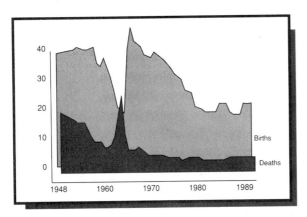

Sources: United Nations 1988, Worldwatch 1989.

In 1977, China began to promote the two-child family by adopting, nationwide, a package of economic incentives and disincentives. The slogan heard everywhere was "one is best, at most two, never a third." Couples that pledged to have only one child received monthly cash payments for 14 years, preferences in housing and job assignments for urban residents, and generous allocations of private land and large rations of grain for rural residents. Single children received priority in medical care, schooling, and jobs.

Couples that had a third child received severe penalties. They had to accept reduced salaries for 14 years, received low priority status in housing and employment, and were made to pay their childrens' cost of medical care and schooling.

The harshness of these measures has caused some resentment among Chinese families, particularly those in rural areas, where large families provide a form of social security to the elderly. Similar resentments were felt in India when the Indian government embarked upon an overzealous program of compulsory sterilization in the 1970s. Voluntary programs are usually more effective than coercion because they work with, rather than against, cultural traditions and don't stimulate animosity toward the program.

Between 1970 and 1980, Indonesia established more than 40,000 family planning centers that provided free contraceptives and information. By 1989, more than 18.6 million families were using birth control and infant mortality had fallen by 40 percent. The fertility rate dropped from 5.6 to 3.4 children per family. While the Indonesian program has moved more slowly than the Chinese program, it enjoys widespread support and can be expected to accomplish a stable rate of births and deaths early in the 21st century.

The population of the world passed 5 billion some time in 1987 and is now increasing nearly 10 percent every 5 years. However, the absolute rate of increase has begun to slow. By 2020 or 2025, it may have fallen to half its present rate, to between 4 and 5 percent every 5 years. However, because the rate of increase applies to an ever-larger population, the total number of people born each year will continue to increase for several more decades. It will be well into the 21st century before the number of new people added each year returns to the already very high levels of today.

If the entire world could achieve replacement level fertility—1 to 2 children per couple—by the year 2010, global population could be stabilized at 7.7 billion by 2060. If we cannot reach replacement level until 2035, population will reach 10.2 billion by 2095. If it takes the full effects of global climate change to spur us to attain zero population growth, and we don't attain that until 2065, global population in 2100 will be over 14 billion.

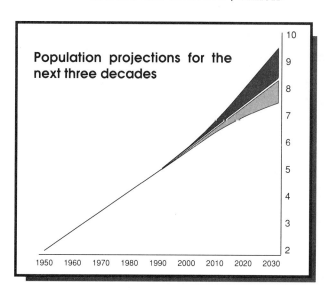

Population projections for the next three decades

If population growth rates were brought to 1 percent per year, a 75 percent reduction, the total number of people in the world would reach 8.5 billion in 2025. Compulsory birth control measures could reduce population growth to slightly more than one half percent annually, resulting in a population of 7.6 billion by 2025. With no birth control, growth rates are expected to climb to nearly 2 percent per year by the year 2000, resulting in a total population of 9.4 billion by 2025.

Sources: Kayfitz, 1989; United Nations, 1988.

At a world population of 14 billion, or even 10 billion, our options will be considerably fewer. Personal liberty will be more curtailed. The demand for food will have become paramount. Tensions will be much higher. A reduced standard of living, mass unemployment, homelessness and hunger all lie along the path of overpopulation. If we want to avoid these punishments, we must find ways to limit growth.

Leadership must come from those that are most prosperous. It is in wealthy nations that people are most able to ask whether the effort of raising another child is worthwhile to them or whether they would prefer pursuing other goals. By associating small families with prosperity, privileged nations can provide an example for the rest of the world.

2. EQUALIZE EDUCATIONAL OPPORTUNITIES FOR MEN AND WOMEN.

Across a wide spectrum of cultures, one relationship emerges: the more schooling women achieve, the fewer children they choose to bear. In Bangladesh, Senegal and Uganda, barely half of all school-age girls attend school. Educational systems have been overwhelmed by the numbers of children, and it has been the girls who have been turned away. It costs about $50 per year to educate a child—$6 billion would educate all the children of lesser developed countries this year. Another $2 billion this year would provide remedial literacy classes for women beyond school age. We cannot hope to defuse the population bomb without broadening the opportunities for women. Education is where that process begins.

The world as a whole must seek to raise the status of women to a parity with men, particularly as to how family planning decisions are made.

3. IMPROVE HEALTH CARE SERVICES TO THE POOR.

A greater sustained level of prenatal, neonatal, and infant health care is needed in the developing world. No technology exported from the industrial world is more important than this simple medical service.

There are those who may argue that providing maternal and pediatric medical services to poor women and infants will mean that more children will survive to reproduce and that women's reproductive lives will be extended. This may be true in the near term, but it is not a valid criticism. *The expectation that a child will live to adulthood is a necessary prerequisite to the voluntary decision to have only one or two children.* The decision to have smaller families is influenced by other factors, some of them taking longer to establish themselves across cultures, some of them brought about by very subtle changes in attitude. But if this prerequisite is ignored, those other influences won't matter. Infant mortality must be reduced and establishing better primary health care systems is the way to accomplish that.

The critical elements of primary health care are relatively inexpensive by medical and development standards: an organized system of trained birth attendants and village health care workers; protection against tetanus, diphtheria, measles, polio, and tuberculosis; improvements in water quality, sanitation, and nutrition; training mothers in oral rehydration for diarrhea; supplemental food and vitamins for pregnant women and nursing mothers; and health and fertility education clinics.

4. MAKE CITIES MORE LIVABLE.

In addressing the problem of overpopulation, we should focus substantial efforts at improvements in the major cities. Every crisis in the environment produces refugees who migrate towards cities in search of a new beginning. In the coming era of multiplying catastrophes, we can expect to see more environmental refugees than ever before. In 1950, less than one person out of five lived in a city. By the year 2020, half the population of the world will live in cities. Unless the cities are able to absorb the overflow and house and employ people in vertical configurations, the tendency will be for them to spread out horizontally. Only by increasing the density of cities can we save wilderness. We need to protect our natural areas because they are the ultimate sources of our life support. We must therefore concentrate our efforts at making cities more habitable for more people in the near term.

Cities typically devote almost all available space to buildings and streets. That is an impoverishing choice. Because of the short distances between living, work, and other locations, cities are ideal places to eliminate petroleum-based personal vehicles and to substitute bicycles, electric cars, and underground mass-transit. A shift away from cars would allow streets to be torn up and replaced with parks, lawns, and gardens, traversed by walking trails and bike paths.

In recent years Jakarta, Indonesia banned pedal-powered rickshaws and then confiscated and destroyed more than 100,000 of them. At the same time, cities like Davis, California and Boulder, Colorado established dedicated bike routes for commuters and in the process improved air quality and reduced levels of noise and soot. Jakarta wanted to rid itself of a stereotype it felt was demeaning. As a result, it condemned itself to becoming a more polluted, more costly, and less healthy place to live.

Cities are ideal places for people to share large appliances, infrequently-used tools, libraries, and other material enrichments, rather than duplicating and hoarding these resources. Health care, education, and other amenities are much easier to deliver in cities than in the countryside, which is why city-dwellers, on average, live longer than people in rural areas.

Cities are especially well suited to put people to work. The labor force is there. By using public works to redress unemployment, cities can meet the double objective of alleviating poverty and improving the quality of city life. When San Francisco was leveled by earthquake and fire in 1906, the city put its population to work and completely rebuilt itself within three years. The difficulty of renovating a major city is often overstated. The only difficulty is in finding the motivation to begin.

The United States may spend as much as 400 billion dollars to finance a permanent space station and a manned mission to Mars. For the same amount of money it could build a $40,000 home or condominium apartment for each of 10 million people. Instead of spending all that money to send one or two men to Mars, NASA could send an unmanned probe and assemble the returning data into a computer flight simulator—artificially-generated images of the landscape—that would allow anyone on Earth with a home computer to "fly" to Mars and explore the planet. Four hundred billion should be spent to redevelop Mexico City, Dhaka, Cairo, and other population centers. That scale of effort—a "mission to Planet Earth"—could establish a new standard for urban living.

5. Protect Indigenous Peoples.

Developmental and industrialization processes frequently impinge on indigenous or tribal societies that had remained relatively isolated until improved transportation and communication broke through the isolating barriers. These societies exist in North and South America, Northern Scandinavia, Australia, Africa and Asia. Their existence is endangered by development.

While the health, nutrition, and education of these peoples may be poor by the standards of the outside world, their knowledge of living with nature is far richer than most outsiders'. Their very isolation usually signifies a special ability to sustain a close-knit culture within a very restricted or hostile environment.

Growing interaction with the larger world is increasing the vulnerability of these inherently sustainable societies. Instead of being regarded as delicate models in human behavior, they are being dispossessed and marginalized, deprived of their cultures, and even—as for the Cinta Larga, Uru-Eu-Wau-Wau, Arara, Gavião, Zoró, Suruí, and Yanomami—genocidally exterminated.

These societies are the links between humanity and its ancient origins in the natural world. It is ironic that even as we make Herculean efforts to feed ourselves by attempting to farm and ranch in rainforests, desert regions, and in other inhospitable locations, we are destroying the only human cultures that have proved able to live for very long times in these places.

We should also recognize that, sometimes, it is the ancient tradition that is destroying the ecosystem. Where traditional styles of subsistence hunting and nomadic herding involve small withdrawals of resources over large areas of uninhabited lands, these lands can be regenerated only if those impacts are kept within limits. Population pressures tend to squeeze larger numbers into smaller areas and to permanently reduce the supporting resources. To preserve productivity, year after year, something has to give. Either the larger area has to be protected from encroachments, even when it seems to an outside viewpoint to be unused, or the number of those permitted to live a traditional lifestyle must be restricted. If neither of these choices is made, the traditional practices become jeopardized and that special skill and knowledge may be lost to the world.

A vision of sustainable development must include the commitment, whenever and wherever possible, to protect ancient cultures by preserving the integrity of their ecology. Paradoxically, it is unavoidable that improved health care, a better supply of fresh water, equal access to education, and emergency assistance in time of need can be destructive of traditional styles of subsistence living. There is a fine line between creating an artificial, unfair isolation, and destructive assimilation by language, attire, evangelism, television, and various intellectual pollutants. Cultural development that overrides human and environmental considerations of a much larger scope and far longer range is not progress. General recognition of that simple fact can mean the difference between endurance and loss of ancient knowledge, between survival and loss of fragile ecosystems, and ultimately, between continued habitability of the natural world and loss of our planet's favorable climate for life.

ENERGY

The most direct human influence on global climate is the production of energy to meet human needs. After deforestation for crops and cattle production, it is to make firewood and charcoal that most trees are cut. It is to produce electricity, steel and steam that most coal is burned. It is to power automobiles, ships, and airplanes that most oil is consumed.

Carbon dioxide accounts for about half of the global warming now occurring. In order to produce less carbon dioxide, we will be required to use less coal, less oil, and allow less burning in general. Yet our entire human civilization is based on fire. This is no small problem.

There are three principal ways to displace fossil fuels: we can improve energy efficiency (accomplish the same output with less input); we can develop renewable and more benign sources of energy (live on our daily income, not our capital reserves); and we can expand the use of nuclear energy (fission now, fusion later). Of these choices, only the third is an impractical and unethical solution. The first two hold great promise, but much needs to be done to bring that promise to fruition.

6. PHASE OUT FUNDING FOR NUCLEAR FISSION AND FUSION.

Nuclear energy is not a viable option because it is not cost-effective. It costs well over a billion dollars to build even a small nuclear reactor. As a heat source, nuclear energy is equivalent to buying oil at $200 per barrel, more than 10 times current prices. No electric utility in its right mind is going to pay 10 times more than it has to to generate its product. The nuclear power plants that have been built to date were all built with massive government and international aid subsidies—several trillion dollars when all costs are summed. It is as yet uncertain whether the energy produced is equal to the energy consumed in the effort.

Nuclear fusion is equally, if not more, expensive. Apart from the uncertain prospect of cold fusion, most fusion processes presently envisioned will be tremendously capital intensive. The research funding alone, which must be sustained over many decades to come, could bankrupt the energy budgets of nations that decide to pursue high-temperature fusion. Even if an international consortium decided to embark on that journey, there is no promise that fusion can ever become a practical source of commercial electricity.

Because of the high cost, any money spent to create more nuclear electricity is money that will not be spent in other, more immediately productive, pursuits. For a nation like the United States to cut carbon dioxide emissions by 50 percent using nuclear-fission power plants would cost at least 50 trillion dollars and require the installation of one nuclear plant every two and one-half days for 38 years. It simply can't be done.

Another reason that nuclear energy is not a viable option is that it is unsafe. One Chernobyl-scale accident every five or ten years is an inevitable consequence of light-water reactors in the hands of light-headed men. Inevitable. Even without accidents, some 250 radioactive isotopes are released into the environment by the fissioning of uranium fuels. These invisible radionuclides are inhaled, ingested, swallowed, absorbed, and passed around the biological domain we inhabit. Cancer and birth defects, diseases of the immune system, and general increases in the ill health of populations living near nuclear reactors and nuclear waste dumps are steadily increasing as the pollution from commercial fission spreads.

Finally, nuclear energy is not a viable option because any nation with a nuclear reactor is a nation with the ability to construct nuclear weapons. The fissioning of uranium fuel creates plutonium. A handful of plutonium, evenly compressed and imploded under the force of conventional high-explosives, can eliminate a city. A single nuclear reactor contains the fissionable material of 1000 Hiroshima bombs. Legal commerce in nuclear fuels creates opportunities for state paramilitaries, terrorists, and revolutionaries to build bombs. One day the world may wake up and find London, New York, Montreal, or Beirut gone and not even know who was responsible.

Increasing ionizing radiation in millirems per year average individual dose from all sources 1949-1989

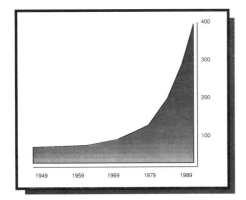

Sources: Boardman, 1989; Nuclear Regulatory Commission, 1989.

Today, even the most powerful nations are powerless to free hostages held by numerous small religious sects in the Mideast. What happens when a local terrorist organization becomes able to hold an entire nation hostage with fuel or waste stolen from a nuclear reactor?

Someday, the superpowers may come to their senses and recognize that only a tiny fraction of the nuclear weapons in present-day arsenals, if used in defense, or aggression, or heat of the moment, would trigger a nuclear winter from which no living being would emerge: no winners, no losers, no witnesses. That day, they may decide to scale back their stockpiles to a point collectively below the nuclear winter threshold, or even eliminate them entirely. Japan eliminated pistols, rifles and cannon for more than a century, and might have continued to do without such weapons had its port cities not been shelled by the American Fleet in 1854. Although nuclear weapons can be eliminated temporarily, the knowledge of how to build them cannot. It is available in many encyclopedias. What can be removed, however, is the panic that pushes the button. Nuclear bomb factories can be dismantled, meaning that before a nation could employ nuclear weapons, it would have to build the factories and then the weapons. The extra time that takes may be time enough for reason to prevail. Sixty days, or even six days, is more time than six minutes, or sixty seconds.

As long as nuclear reactors are spread around the world, the time it takes to build nuclear bombs is reduced to a matter of hours. Nuclear reactors are bomb factories. Nuclear reactors threaten humanity with annihilation just as much as do stockpiles of nuclear weapons. Nuclear energy and long-term human survival are mutually exclusive.

National security is the Achilles heel of any centralized, capital intensive, and complex energy system. By accident or act of malice an entire system and all its tendrils into industry, transportation, government, and public health can be disrupted in a few seconds, throwing interlinked infrastructures of entire regions into chaos. True security for individuals, and for nations, derives from using resilient, low technology, free-standing, and dispersed sources, not only for energy, but for all the requisites of life.

7. ENCOURAGE ENERGY EFFICIENCY.

Of the two remaining options, energy efficiency and renewables, the most cost-effective is energy efficiency. It is also the most immediately available carbon-replacing option for large scale use.

A new coal-fired power plant produces power at slightly less than 2 cents per kilowatt hour. Nuclear planners project the day, far in the future, when improvements in design and construction will bring the price of nuclear electricity from 13.5 cents per kilowatt hour (subsidized) down to 5 cents per kilowatt hour (subsidized). Today, improvements of energy efficiency can be purchased at an average cost of 2 cents per kilowatt hour and the hope is that in the very near future, investments in research will bring that price down to *half a cent* per kilowatt hour.

**World energy growth
in horsepower and electricity
1900-1988**

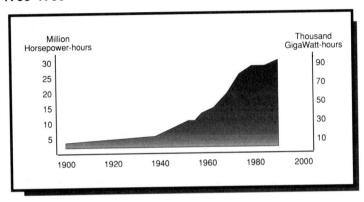

Source: Gibbons, et al., 1989.

Energy efficiencies can be achieved in both production of electricity and in the way it is distributed and used. A rigorous program of power plant rehabilitation could decrease the amount of fossil fuel consumed in the production of electricity by as much as 10 percent. Retrofits such as atmospheric or pressurized fluidized bed combustion, integrated gasification/combined cycle, and cogeneration can more than double that increase in production efficiency. Advanced technologies such as magnetohydrodynamics (passing hot combustion gases through a magnetic field to create electricity) promise to more than double the efficiency of even fluid-bed power plants.

A good example of end-use efficiency is a new, screw-in fluorescent light bulb that produces the same illumination as a 100-watt incandescent bulb but draws only 24 watts of power. Each of these bulbs that is installed saves 400 pounds of coal from being burned. If all of the incandescent light bulbs in the United States were replaced with fluorescent light bulbs, the energy savings would be equivalent to construction of 40 large power plants and about $10 billion dollars each year in reduced fuel costs.

Energy efficient light bulbs

Light Bulb	Output (Lumens)	Power (Watts)	Efficiency (% output per watt)	Life (Hours)
25W Incandescent	235	25	11%	1250
PL5 Fluorescent	250	5	63%	10000
40W Incandescent	455	40	14%	1000
PL9 Fluorescent	575	9	77%	10000
60W Incandescent	860	60	17%	1000
PL13 Fluorescent	900	13	82%	10000
75W Incandescent	1180	75	19%	750
PL18 Fluorescent	1250	18	83%	12000
100W Incandescent	1750	100	21%	750
PL24 Fluorescent	1800	24	90%	12000

Developed by the federal grant programs in the 1970s, the compact fluorescent bulb operates by passing a 100 volt electric current (provided by a transistorized "ballast") through a gas-filled glass tube, generating invisible ultraviolet radiation. A phosphorescent coating on the inside of the tube captures the ultraviolet light and glows white. These bulbs can be ordered from United Helping Hand, a charity for the handicapped (in the United States telephone 1-800-877-2852).

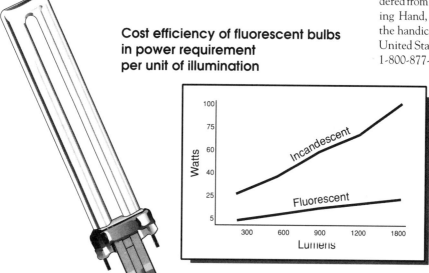

Cost efficiency of fluorescent bulbs in power requirement per unit of illumination

Source: Dankoff, 1989.

Between 1973 and 1987, the efficiency of buildings, industry, and transportation in the United States improved 26 percent with only minimal government assistance and, in some cases, considerable government resistance. This improvement reduced U.S. carbon emissions by 400 million tons per year, an amount roughly equal to the carbon produced by all the world's automobiles.

8. IMPROVE TRANSPORTATION SYSTEMS.

The average automobile produces several times its own weight in carbon each year, which it pumps into the atmosphere as carbon monoxide and carbon dioxide. By increasing the number of miles that each automobile travels on a gallon of gas, we reduce those carbon emissions. Beginning in the 1970s, a number of countries enacted fuel conservation standards in order to reduce petroleum imports. These standards dramatically reduced the operating costs of new cars, and gradually improved the efficiency of automotive technology in general.

If United States gasoline mileage standards were increased once more, from 19 miles per gallon (present fleet average) to 26 miles per gallon, it would save the equivalent of an Alaskan oil pipeline in petroleum (almost the entire supply of oil from Alaska) each year, and yield $220 billion per year in savings to consumers. If world fuel performance were increased to 50 miles per gallon, world carbon emissions from transportation could be cut in half, even if the world fleet increased to 500 million cars. Using only existing technology, most of the world's car manufacturers could achieve a new-car average of from 50 to 80 mpg by the year 2000 while maintaining or improving current standards of safety, air pollutant emissions, comfort, and performance.

Average gasoline vehicle efficiency for the United States, in miles per gallon 1970-1988

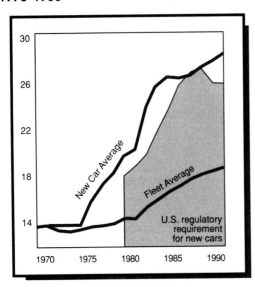

Sources: U.S. Motor Vehicle Manufacturers Association, 1988; Worldwatch Institute, 1989.

Solar racing car

General Motors' Sunraycer won the first World Solar Challenge race across Australia in 1987 and set eight international electric car speed records in 1988. It has been donated to the Smithsonian Institution.

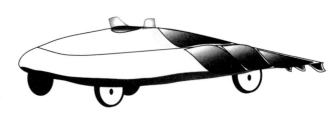

In the future, automobile designs will change dramatically from what they are today. Already, solar-powered cars are setting electric vehicle speed records in races across continents. In 40 years time, there may be no fewer than 3 million solar powered cars in the United States, and 8 million in Europe. One manufacturer, the Solar Car Corporation of Melbourne, Florida, is currently testing commuter vans that can carry normal loads at 55 miles per hour (88.5 kph).

**Available gasoline vehicle efficiency
4-passenger automobiles
currently in development or production
in miles per gallon**

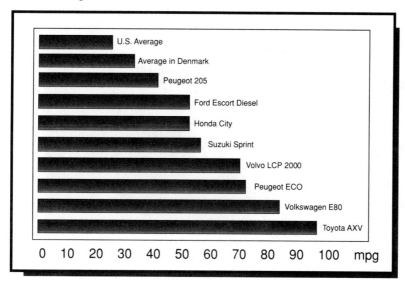

Sources: Bleviss, 1989; Worldwatch Institute, 1989.

Solar-wind hybrid electric vehicle

The Amick WindMobile uses an arch as an airfoil to propel the vehicle at 5 times windspeed without batteries or 3 times windspeed with batteries and solar cells. The car, which has been operated since 1976, can reach speeds of 70 miles per hour.

Hydrogen fuels, developed for the space flight program, could soon find their way to the corner gas station. Daimler-Benz of West Germany and Billings Energy Corporation of Provo, Utah, are currently operating test vehicles. With reduced drag and weight, passenger vehicles twenty years from now could get 150 to 200 miles per gallon equivalent, at a price of perhaps $1.35 per gallon (in 1990 dollars). Production of fuels and prototype vehicles is already underway. Early in the next century, Japan may become a net energy exporter. Its product: hydrogen fuels from seawater.

Solar commuter car

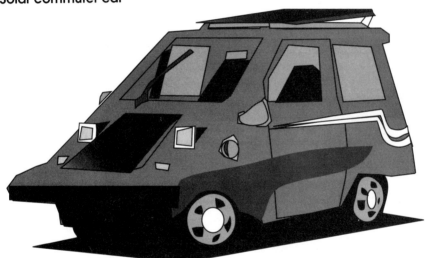

Solar Car Corporation's prototype vehicle trades longer range and faster speed for increased safety, comfort, and carrying capacity. The vehicle uses standard tires, brakes and suspension and an aluminum strut roll cage surrounds the passenger compartment. It can reach 55 mph with a range of 20 to 30 miles on a sunny day, and can be recharged externally for longer distances.

The average citizen in an industrialized nation eats more food, owns more clothes, and has more expensive entertainment than the average person in a less developed nation, but these differences are not substantial from an environmental standpoint. A much larger disparity exists in the amount and means of travel—as commuters, vacationers, or for business purposes. Americans see nothing wrong with flying to the Bahamas to go swimming, or driving 1,000 miles to visit a national park. Indeed, most people aspire to have the time and money to be able to do that more often. The amount of carbon dioxide and other greenhouse gases produced in travel-related activity is large and growing larger. The luxury of mobility costs much more than is now apparent, but as environmental costs come to be factored in, such as through a carbon dioxide tax, alternative forms of transportation will begin to attract more attention.

Transportation Options

- Automobile weight reduction, aerodynamic and drag improvement, engine and drive train improvement
- Light-duty vehicles, commuter vehicles, and short-distance vehicles which employ smaller engines, human power, solar power, and/or alternate fuels
- Adiabatic diesel engines that enable large trucks to get 80 mpg
- Convertible boxcars that are interchangeable as truck-trailers and railcars
- Wind-assisted and advanced diesel cargo ships
- Fuel-efficient and alternate-fuel aircraft
- Development of more efficient gasoline fuels, hydrogen and alcohol fuels, specialized fuel cells, and photovoltaics
- Encouragement of urban mass transit and high-speed and magnetic-levitation rail systems
- Expansion of telecommunications
- Increased availability of bicycle lanes and special-purpose roadways

The most efficient means of transportation, in energy conversion terms, is human power. While most urban and rural residents live within easy bicycling distance to work, only in non-industrial countries are bicycles widely used as commuter vehicles. There is a growing willingness in Western cities to bicycle to work, for reasons of health and enjoyment as well as economy, but there are also strong disincentives. Where bicycles share the same roadbed as gas-powered vehicles, the bicyclist is at a disadvantage. His life is at risk. When he

Human powered vehicle

Lightweight, safe, and comfortable pedal-powered vehicles could find their way into many cities in the future. This prototype, designed at the California State University at Chico, is capable of speeds of 45 miles per hour.

arrives at work, sweating from the aerobic exercise, he may have no way to shower or change his attire. These problems are easy to solve and require no great sacrifice. All that is needed are dedicated bike trails, bike parking areas, and shower houses and locker rooms. The benefits, in elimination of urban pollution, reduction of fossil fuel consumption, and improvement of health, are easily worth the cost.

In a few years it should be possible for a family to vacation at the mountains or the shore by taking light rail transit and bicycling locally. The world will be crisscrossed by interlocking bike-rail terminals and bike trails that enable people to travel from city to city and country to country efficiently, economically, and ecologically.

9. REDESIGN BUILDINGS

Recent improvements in North American building design have spared the atmosphere 225 million tons of carbon annually, but the buildings that are unimproved still contribute 900 million tons of CO_2—17 percent of world carbon emissions. We could displace 85 coal-fired power plants, or 2 Alaskan oil pipelines, with only small design changes in office building construction—*at no additional cost!*

In most cities of the Northern Hemisphere today, acres of floor-to-ceiling single-glazed windows leak warmth to the outside in December and soak up heat from the summer sun in July. Add to these windows of vulnerability a wide array of inefficient machines, lights, and room circulation patterns—inherited from an era of cheap oil—and it's no wonder that 5 to 10 percent of total floor space has to be devoted to air conditioning equipment. Over each of these buildings' lifespans they will require two, or even three, times as much as it cost to build them just to keep them at habitable temperature.

Solar Architecture

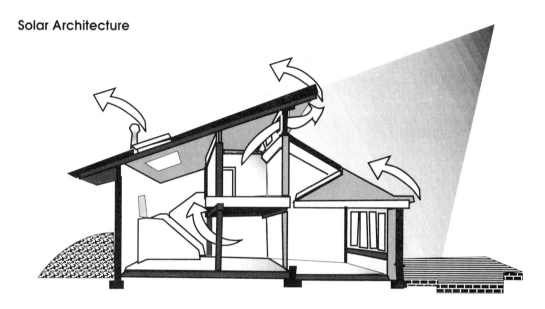

Passive solar design, used by the ancient Greeks, employs sunlight and thermal wind currents to heat and cool the home. Solar radiation is collected directly from sun and sky, and by reflection from terrain. The heat is stored in the thick masonry of the building. Earth berm on the shaded side keeps the house warmer in the winter and cooler in the summer. Glass and terraces or reflecting pools on the sunlit side increase the solar intake, but can be shaded from the higher summer sun by plants or overhangs. Ventilation allows heat to circulate from warm areas to cool areas without obstruction, keeping the entire building at a constant temperature.

Source: Schecter, 1989.

Just as commercial building designs have been revolutionized in an era of costlier fuels, so have residential designs. In the United States, houses built with superinsulated walls and ceilings, ventilation systems that recover heat, special doors and windows, and more efficient furnaces reduce home fuel needs by more than 75 percent. These buildings take advantage of the "free" heat from people, lighting, appliances, and sunlight. Tapping these unused sources can raise indoor temperatures by 30°F (17°C). If the thermostat is set at 70°F (21°C), the furnace is not needed until outside temperature drops below 40°F (4°C).

In the late 1970s the United States Congress allocated $16 million to develop new technology to use energy more efficiency. Among the achievements by laboratories that received funding were breakthrough designs for compact fluorescent light bulbs, window coatings that retain heat inside buildings, and new, low-cost insulating materials. When fully implemented, the $16 million invested in research is expected to return $64 billion in energy savings, *a 400 million percent return.*

Superinsulating Windows

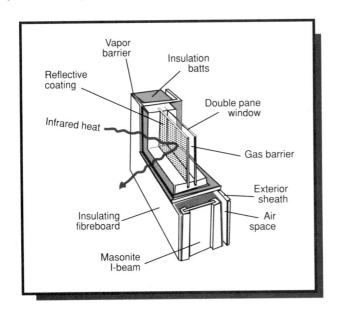

This window design was developed in Sweden but is now commercially available in many countries. By using inexpensive building techniques and materials, a window can be as well insulated as a standard wall (R-11 or better), saving more than 75 percent of winter heating costs.

Source: Rosenfeld and Hafemeister, 1988.

10. EXPAND FUNDING FOR SUPERCONDUCTIVITY.

Superconductors and supermagnets offer little or no resistance to electrical flow. Recent breakthroughs in low-temperature superconductors made from exotic alloys offer the promise of far greater efficiencies in generation, transmission and storage of electricity. New iron alloys which are not superconducting also have improved magnetic strengths. These metals align their crystal lattice in one direction, which increases the magnetic pull of the alloy.

Reduction in size and weight of motors using supermagnetic alloys

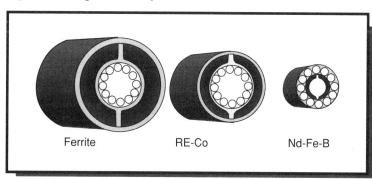

Motors using advanced alloys of cobalt (RE-Co) and neodymium (Nd-Fe-B) promise significant reductions in size and weight. A neodymium motor powered the GM Sunraycer.

Source: Cartoceti, 1984.

By making much more powerful electromagnetic fields, superconductors and supermagnets will greatly decrease the amount of fuel required to produce electricity. They will greatly diminish the weight, size and power requirements of electric motors used in homes, industry and transportation. By reducing or eliminating transmission line losses, they could save electricity as it is distributed rather than wasting 8 percent as resistance heat. By being able to store energy with efficiencies of 95 percent, compared to 75 percent for pumped storage, or 65 percent for battery storage, superconductors and supermagnets will increase the practicality of photovoltaic, wind, hydro, and other renewable but variable sources, and dramatically expand the power and range of electric and alternate fuel vehicles. Superconductivity is still in development, but the promise that it holds is very great.

11. USE SOLAR POWER.

In the next 100 years, the human population will greatly expand its consumption of electricity. New sources of hydroelectric power are few. If we use nuclear energy, we will poison ourselves. If this electricity comes from coal or oil, it will destroy the planet. But there are other sources.

Photovoltaics (energy from sunlight) is now a realistic alternative for large-scale energy production. The primary element used in photovoltaic crystals is silicon—no more than refined sand. Silicon makes up a large part of the Earth's crust. Advances in the next two decades could make photovoltaics the first choice for new electrical production.

Because of their expense, photovoltaic systems were not considered suitable for large-scale electrical generation until after the 1973 oil crisis. An international research effort has since set new performance standards for crystalline arrays. Inexpensive semicrystaline and polyamorphous arrays have been developed. The cost of photovoltaic power has dropped from $15 to $0.30 per kilowatt hour in ten years and is expected to decline another 80 percent by the year 2000.

For nearly ten years, one photovoltaic factory in North America has been building modules of new photovoltaic cells by using the power of sunlight captured on its own roof. The sunlight falling on virtually any unshaded rooftop is more than enough to power a modern home equipped with all the usual appliances enjoyed by the wealthy in the industrialized world: washing machines, refrigerators, televisions, and computers. Photovoltaics produce no carbon dioxide, methane, or other greenhouse gases. They are an environmentally safe way to raise the average standard of living in most countries of the world.

Photovoltaic prices and shipments worldwide, 1975-1988 in dollars per peak watt average and megawatts installed

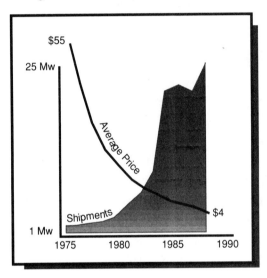

Source: Worldwatch, 1988

Photovoltaic cell

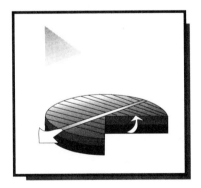

A photovoltaic cell is made up of a negatively charged upper layer and a positively charged lower layer, separated by a junction. The cell accepts photons from the sun and dislodges electrons which are collected by a wire grid and transported away to be used as electricity.

For years, photovoltaic research has been an orphan. Government energy research programs spent billions on nuclear fusion and fission and tossed pennies to alternatives. Yet to make photovoltaics competitive with carbon-producing sources, governments must play an important role in helping to develop the industry. Some countries are beginning to do this. It needs to happen everywhere.

12. Choose renewable sources.

Other renewable sources are also promising, but depend to a larger extent upon regional and seasonal resources. In the past five years, the demand for wind turbines has risen ten-fold in China, Denmark, Greece, India, and Spain. In the summer of 1987, wind power accounted for 5 percent of the electricity sold by Pacific Gas and Electric, one of North America's largest utilities.

Sweden is focusing its efforts on biomass—wood, energy crops, and agricultural surpluses. While biomass sources still produce carbon dioxide and other greenhouse gases, the production is at a steady state with consumption: growing plants absorb carbon from the atmosphere. The substitution of wood fuels (methanol for cars, for instance) for fossil fuels throughout the world could permanently reduce emissions of carbon to the atmosphere by 6 to 8 billion tons per year.

With government development dollars, geothermal energy— heat from the Earth's core—has grown to 5 gigawatts (5 million kilowatts—a medium city) of capacity worldwide in the past few years. Another 2 gigawatts will enter service by 1991. North America alone has the unused potential to produce as much as 18.7 gigawatts of geothermal power by the year 2000, and as technology improves, that potential will increase.

Other sustainable energy systems are coming along rapidly: methanol and ethanol fuels, solar water heating, passive solar space heating, solar ponds, large solar concentrators and furnaces, solar-powered Sterling and Rankine engines, tidal power, ocean thermal gradients and wave motion systems, microhydro, and cogeneration. All of these will make significant contributions in the future.

13. Establish world targets.

Through energy efficiency improvements, the U.S. economy improved 40 percent in the past 15 years with no increase in total energy consumption. Still, Japan uses 36 percent less energy to produce a dollar's worth of manufactured goods than does the United States, and that advantage is widening. The United States has the means to reduce its energy costs by $220 billion per year, above and beyond the $150 billion already saved by recent improvements. The price for this cost reduction would be an investment of only $50 billion, meaning that the investment could be completely repaid after just 83 days.

The world could have a 2 percent annual rate of efficiency improvement for the next decade without having to come up with any new technology—we already know how to do it. To sustain a 2 percent improvement rate beyond the year 2000 will require a little more effort and innovation, but that effort, estimated to cost $15 billion, would reduce energy consumption to one-third of present levels over a 50-year period.

Some countries have been achieving this 2 percent target for the past 15 years. The technology is steadily improving. We can make buildings all over the world as efficient as they are in Sweden, industrial factories as efficient as they are in Japan, and automobiles as efficient as the best French and German prototypes, with very little additional capital outlay, and with a rapid pay-back from reduced energy overhead. Why don't we do it? Up until now, we haven't considered it that important.

In the United States, the average citizen has 200 energy slaves. That is to say, in human equivalent power, each American's appliances provide an average of 200 times the total power level of human metabolism. These slaves consume mostly fossil fuel, burned at the rate of 8.75 billion joules per day, or 10.5 kilowatts per capita. In the 27 next richest countries of the world, the average citizen has 120 energy slaves, consuming about 6.2 kilowatts per capita. The world average is about 40 energy slaves, although in the poorest parts of the world, there are no energy slaves at all. Some of the world still exists as it did before electricity.

If we view the world from the perspective of nature, humans have endowed themselves with far more than they need for mere survival. Ironically, they have gone so far beyond their needs that their profligacy now threatens their survival.

We don't have to free our energy slaves. What we must do is to provide for those slaves—the appliances and conveniences that make life easier—from our daily income from the sun, rather than from what we can beg from the Earth and steal from our children.

14. TAX GREENHOUSE GASES.

Much of the debate in international forums of the past two decades has been over the philosophy of market regulation. On one side are economic purists who see in unfettered markets a natural balance of supply and demand, production and consumption, need and fulfillment. On the other are those who regard free markets as fundamentally unbalanced—taking from the have-nots and further enriching the haves, ignoring responsibilities to the global commons—and ultimately unsustainable.

The economic marketplace of the world today, regardless how free or how regulated any given sector may be, is a rigged market. Energy is too cheap. Natural resources are too cheap. Manufactured goods are too cheap. Few, if any, commodities reflect true costs.

The moment we put a dollar per ton tax on the emission of carbon dioxide, we will see a change in the rate at which industries discharge carbon. It may take a year or two, but everything that can be done to save that dollar, if it costs less than a dollar, will be done. When we reach that point, we need to raise the tax to 2 dollars per ton. Then it needs to go to 3 dollars per ton. Worldwatch Institute has estimated that a carbon tax of $50 per ton would raise the price of gasoline by seventeen cents per gallon and of electricity by 2 cents per kilowatt-hour. It would cost consumers some $240 per year in the United States and $9 per year in India. These are not insurmountable sums, especially if they were reached gradually, over the course of several years. And yet, a $50 tax on each ton of CO_2 could raise nearly $300 billion annually worldwide, more than enough to offset the diverse economic impacts that may accompany the worldwide shift toward sustainability.

Every nation needs to put a cost on destruction of forests, topsoil, and biological diversity. Every nation needs to make pollution prevention pay. But nations are afraid to act unilaterally, because to do so can place them at a competitive disadvantage. The solution is to establish world targets and provide fair incentives to move toward them.

One way to link economic rewards to responsible climate policy would be to tie the international rate of monetary exchange to protection of the environment. Nations could receive credit for strict regulation of manufacturing processes and lose credit for wastefulness and pollution. Currencies could receive value for reductions in birth rates, cessations in production of greenhouse gases, and topsoil preservation. They could lose value for use of high-sulfur coal, loss of forests, or production of acid rain.

We are living wastefully, but we have been insulated from the consequences because we have begun to spend Earth's capital as if it were income. We may not notice that we are borrowing from our children. Our children will notice.

FOOD SECURITY

More than a billion people in the world will go to bed hungry tonight. The number is growing. Ironically, many of these people are farmers. Many live on or near land that was once cropped but is now laying fallow. Many of the countries that are not growing enough food to feed themselves possess the largest reserves of untapped farmland.

15. RESTRUCTURE AGRICULTURAL SUBSIDIES AND INTERNATIONAL AID.

The food surpluses created in North America and Europe over the last two decades were a result of government subsidies and other incentives that stimulated production artificially above world demand. Price supports in the United States alone grew from $2.7 billion in 1980 to $25.8 billion in 1986. These price supports hurt the farmers who grew smaller or unsupported crops, and they hurt the national economy by contributing to enormous deficits. In Europe, the increase was from $6.2 billion in 1976 to $21.5 billion in 1986. The same adverse economic effects ensued.

The United States has traditionally paid for much of its military and other expenses by exporting agricultural surpluses. It also gives away vast quantities of grain in lieu of direct financial aid for Third World development. While the food aid is viewed as charity in the United States and elsewhere, and tallied in dollars, it is actually counterproductive in several respects. Gifts of food drive down prices on the world markets and create severe problems for developing countries whose only significant source of revenue is agricultural produce.

Because most people in the developing world are marginal family farmers to begin with, importing cheaper food from abroad or receiving donations of surplus food as development aid creates unemployment. In the industrial countries, marginal family farmers are also driven out of business as subsidies favor the specific crops that are in demand for export and can be produced best on a larger scale—such as wheat, corn, tobacco, beef, and milk. Loss of family farming ushers in chemical-intensive, machine-intensive, capital-intensive farming that is less in touch with the sustainability of the land.

To reverse these trends, the challenge in the United States and elsewhere is to carefully undo what has been so carefully constructed. Food strategies must shift production closer to where it is most needed, secure the livelihoods of the rural poor, and conserve small farms. The industrial world must reduce subsidies that are not related to resource conservation, eliminate trade structures that confer unequal economic advantages, and reassess the distribution of non-emergency food aid.

There are those who advocate cutting off even emergency food aid to developing countries. They argue that each country must learn to support its population within the confines of its own borders and must suffer the consequences of exceeding finite resources. That is an unconscionable position. Most of the famines in the world today are brought about by politics, international trade imbalances, and unexpected events. The victims are usually peoples who had been self-sufficient farmers and herders for centuries, but who are caught in a disaster not of their own making. While exceptions to this generalization exist, they are rare. If the industrialized world condemns innocent people to starve while it gorges itself, wastes food, and throws the scraps to its animals, it becomes contemptible. Instead, the First World must do what it can to redress the political and social imbalances that bring about Third World famines. It must feed the poorest of the poor and nurse them back to self-sufficiency.

16. PROMOTE SUSTAINABLE AGRICULTURE.

In traditional agriculture, local organic material—cover crops, compost and manure—provided farms with sources of energy, soil nutrients, and ways of controlling pests. In our rush to stimulate agricultural productivity, we have substituted electricity, petroleum products, chemical fertilizers, and pesticides. These substitutes cost much more, provide diminishing returns, create waste, and do lasting ecological damage to the resources upon which succeeding years' crops will depend. A sustainable vision of agriculture seeks to restore the rate at which food is grown to the amount of food that healthy, well-tended soils can continuously regenerate.

World cropland per capita

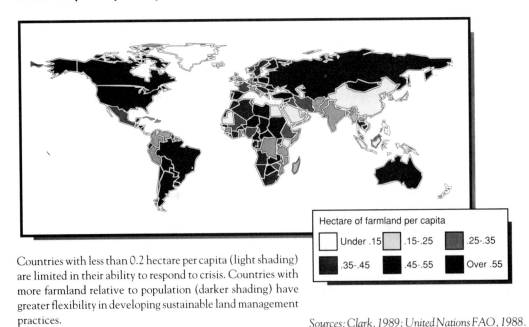

Countries with less than 0.2 hectare per capita (light shading) are limited in their ability to respond to crisis. Countries with more farmland relative to population (darker shading) have greater flexibility in developing sustainable land management practices.

Sources: Clark, 1989; United Nations FAO, 1988.

One of the most important prerequisites to protecting cropland from overproduction is the preservation of enough cropland. Nations such as Japan have already taken the lead in enacting laws that forbid conversion of farmland to nonfarm uses. In 1968, the Japanese recognized that once farmland has been converted to roads, villages, factories or shopping malls, it cannot very easily be restored to food production. Strict Japanese zoning laws now make it illegal to build on cropland.

Smaller scale farms are more productive, acre for acre, than larger agribusinesses. By switching from wasteful irrigation ditches to more efficient systems of drip irrigation, Israel greatly expanded its irrigated areas and boosted the yield of production from each irrigated acre. By encouraging more labor-intensive crops like chickpeas, millet and cassava, many equatorial nations helped subsistence farmers to re-establish themselves in the margins between large plantations. The net effect of helping small farmers has been to bring about an increase in world food production, at a sustainable rate of supply.

17. REDUCE LIVESTOCK PRODUCTION.

When world population reaches 6 billion people, human food needs will begin to affect how much grain is being fed to livestock. In 1989, nearly one third of the grain produced in the world was consumed by domestic animals. The average family dog in America will cost its owners $5,000 during its lifetime, many times more more than the average Chinese farmer will live on in the same period. The grain used to produce a quarter-pound hamburger could feed a person for two days. Cows support vast numbers of anaerobic (oxygenless) bacteria which convert otherwise indigestible cellulose into protein and methane. The protein is used by the cows. The methane—80 million tons per year—is belched into the atmosphere. Animal agriculture also accounts for 85 percent of topsoil loss and produces 20 billion pounds of waste every 24 hours.

It would help for people to shift their diets from beef to poultry, but we must begin to think about going farther. We must begin to eat grains directly and not waste valuable protein.

18. Assist small family farms.

In countries where land is unequally distributed, agricultural production will not grow appreciably without land reform. Large estates with wealthy absentee landlords and vast numbers of landless, hungry poor do not peacefully co-exist in close proximity. Space abhors a vacuum. The hungry will fill up the land. Each nation must reform tenancy arrangements, establish land rights, and create subsistence farming opportunities. Only in this way can protection of forests, preservation of topsoil, and an adequate supply of food be assured.

It is a curious irony, but inevitable, that what is needed to create sustainable agriculture, sustainable forestry, and sustainable industry in the future—to close the loop with nature and return to living within the constraints of natural resistance—is to return to operating at a smaller, closer, more intensive scale. Progress has to change direction and move away from industrialization because industrialization is based on replacing jobs with greenhouse-augmenting machinery. Efforts to create full employment should favor people who craft with their hands. We will need more people to hoe, and weave, and gather forest products.

Neither migration toward or away from the countryside is desirable. Our ability to produce food ecologically is more than sufficient for those who now live in the rural countryside to supply the needs of those who now live in the cities, within limits, without everyone having to leave the cities and grow their own food.

Sustainable agriculture, the only realistic insurance policy for food supply, requires a holistic approach to ecosystems at the local, regional, and global level. Each farmer must begin to live within the sustainable limits of the water, soil, and climate with which his farm has been naturally provided. With good farming practices, hard work, appropriate technology, and a little luck, almost any farm can become productive with the sunlight and rain that arrive in season.

And that is what it will take to feed the world.

Armaments

Where does the money come from to accomplish the programs we need? We need billions for health care, billions for nutrition, billions for carbon-abatement efforts, billions for measures against sea level rise, and billions for improvements to education, agriculture, land reform and other programs. Where will it all come from?

This year the world will spend nearly one trillion dollars—one million million—on weapons of war. World military spending now surpasses the gross national product of half the world's population.

While three-quarters of the current expenditures for arms are in the industrial world, where a single airplane may cost hundreds of millions of dollars, military expenditures in the developing world have also increased dramatically—more than five-fold since the 1960s. Nearly 20 percent of the external debt of developing countries is now attributable to arms imports. Thirty-nine million men now march in armies, state security forces, and guerrilla movements. Another 16 million are employed in the arms industries and as many as 60 to 80 million people work in ministries of defense, military think tanks, and defense-related institutions.

Current military personnel and defense-related employment in persons per 10,000 population

Israel	415
North Korea	350
Singapore	219
Taiwan	217
Soviet Union	160
Turkey	159
South Korea	137
France	102
Spain	101
Norway	98
Chile	95
Egypt	94
United States	92
Italy	91
Sweden	84
West Germany	81
Malaysia	72
Netherlands	70
United Kingdom	59
Peru	57
Pakistan	57
Austria	53
Thailand	48
Argentina	44
China	37
Brazil	34
South Africa	27
Philippines	26
India	18
Indonesia	16

Source: Worldwatch, 1989.

Why do we persist in this wasteful folly? Are we condemned by our past to view the world with such overwhelming military perspective? Some 120 armed conflicts have been fought since World War II, with 20 million deaths resulting. Since 1980, aggressors have won only 1 out of every 10 wars. Almost 9 out of 10 of the victims have been civilian. Doesn't that say something? Doesn't that tell us that resolution of conflict by force of arms has lost its meaning? War is not only obsolete from a theoretical standpoint—it destroys the wealth and homeland, the security, of the warriors—it is obsolete on practical grounds: military aggression has little or no chance of success.

It might be argued that the reason wars are unsuccessful is because of the build-up of defense forces in even the smallest nations. Exactly the opposite is true. The build-up of standing militaries has invited border conflicts more often than not. Constant military readiness hastens resort to military options when political persuasion fails.

With few exceptions (U.S.-Vietnam, U.S.S.R.-Afghanistan, and Iran-Iraq, for example), conflicts have been short-lived because of the tremendous expense and destructive capability of modern weapons. Still, a succession of very expensive little wars demonstrate a continued, widespread acceptance of the central tenet of power politics: that threats are credible only if it appears they can and will be carried out.

Referring to the nuclear weapons capability of the United States, former National Security Advisor Zbigniew Brzenzinski told the New York Times in 1981 that "a credible willingness to use ... implies occasional use." This is the weak spot in the underbelly of deterrence. Weapons of mass destruction deter against their own use, but only if the threat to use them is credible. If presidential advisors like Brzenzinski are whispering sweet proposals to occasionally cross the nuclear firebreak into the ears of the men who hold the buttons, no one is safe. There is no deterrence, no firebreak, and ultimately, no security.

As Dwight Eisenhower said in 1953, "Every gun that is made, every warship launched, every rocket fired represents, in the final analysis, a theft from those who hunger and are not fed, who are cold and not clothed." The homeless cannot build shelters from rocket fuel and the hungry cannot eat gunpowder. Weapons are the least productive use of resources. If they are not used, the resources that went into them are tied up endlessly, and if they are used, those resources are simply destroyed. No bridges are built. No water systems. No rural health care centers. No elementary schools. In Thailand and Guatemala, the military is the largest landholder. In Egypt, it is the largest farmer. In Turkey, it is the largest manufacturer. In the United States, it is the largest financial institution. And yet, virtually all it produces is waste.

The 1989 defense planning budget of the United States called for spending $1,678 billion between 1990 and 1994. This is only about half of the defense budget that the previous U.S. administration had forecast, but it is still a considerable burden on the United States and world resources. Each of 2,100 light helicopters will cost $17 million. The U.S. Marine Corps' 663 vertical takeoff planes will cost $39 million apiece. 750 fighter planes will weigh in at $87 million each. Each of 899 Trident II missiles will cost $40 million dollars. Each of 211 cargo planes will cost $178 million. Each of 132 B-2 bombers will cost $516 million. Each of 12 SSN-21 submarines will cost $1.6 billion. The cost of keeping all of this equipment operational once it is produced will be in the hundreds of billions annually.

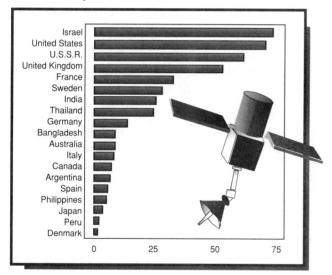

Current expenditures for military research and development as a percent of total public research and development

Sources: OECD, Ulrich Albrecht, Worldwatch 1989.

Weapons research now costs more than the combined expenses for development of energy technologies, improvements in human health, advances in agriculture, and pollution controls. It employs more than one fourth of all the world's scientists and engineers. This is an enormous waste of the precious resources we will need to turn down the global thermostat.

19. TAX MILITARY SPENDING.

The defense of Europe, the amount spent by both NATO and the Warsaw Pact to guard against aggression by the other, is now $600 billion per year. According to studies by the Worldwatch Institute and others, the total needed to turn the world's environmental crisis around—to reduce topsoil loss, reforest the earth, raise energy efficiency, reduce population growth and move towards a sustainable supply of food and water—is only slightly more than one tenth of that amount. About $774 billion will be needed to be spent during the decade of the 1990s. *If $7.74 of every one hundred dollars spent on international arms and militarism were sequestered and the revenue directed to more productive purposes, the climate crisis could be abated.* Seven percent is roughly the same amount as a typical state's sales tax.

In the United States, one third of the entire bill for the Worldwatch plan could be met by what will be spent to clean up the radioactive waste contamination problems at U.S. nuclear weapons production facilities. The Trident II submarine and F-16 jet fighter programs will cost $100 billion, enough to power 30 cities of 200,000 entirely by solar electricity. One Trident submarine costs more than a 5-year immunization program that could prevent a million deaths among children. Really, which is worth more to U.S. security and international prestige?

The Stealth bomber development program—to create an "invisible" plane that may never actually fool anyone—will cost $68 billion, two-thirds of the estimated cost of U.S. clean water goals from now until the year 2000. By foregoing one nuclear weapon test, America, Britain, France, the Soviet Union, China, Israel, South Africa, Pakistan, or India could bring safe water to 80,000 rural villages. One hour of flying a modern bomber on practice missions costs $21,000—more than it would cost to bring maternal and neonatal health care to 10 Third World villages.

What is in the mind of the pilot of a B-2 bomber as he circumnavigates the globe, kept constantly airborne by a 1950s-era plan to maintain war readiness? As he looks down at the vanishing forests, the blowing topsoils, the spreading deserts, does he consider his role in the slow process of the planet's destruction? He must know that a typical mission will cost his government more than he will earn in his entire life. He knows that he is only one of hundreds of pilots in hundreds of planes and that the same mission will be flown again tomorrow, and the next day, and the next. He knows that if he doesn't fly the mission, someone else will.

Maybe, if that pilot is conscientious and thoughtful, he wishes the world would stop wasting money, and his life, in this way. Maybe, he thinks, we could preserve freedom without destroying life on Earth in the process.

REFORESTATION

To slow down and eventually reverse global warming we must reduce greenhouse gases in the atmosphere. The most direct and effective measure that can be undertaken to reduce the greenhouse effect is reforestation.

To adequately counterbalance the emission of carbon dioxide from the burning of fossil fuels at current levels, the world would have to create some 2.7 million square miles (7 million sq. km.) of permanent forest that does not now exist. This is an enormous area of land—roughly the size of Australia if all of it were in one location. Yet, it may not seem as large if we look at what is available in different parts of the world today.

20. REFOREST AVAILABLE LANDS.

When forested lands are cut down and pastures are planted for cattle, it is often the case that the pastures do not remain for very long periods. In some cases, the pastureland is too poor to be sustained. In other cases, the return from beef production is less than the cost of keeping the land in pasture. In much of Latin America, planted pastures support less than 1 cow per 10 acres (0.3 animal per hectare) and, if not replanted, are heavily invaded by brush and trees within 5 or 6 years. More than 386,000 square miles (1 million sq. km.) of such pastures in Latin America are suitable for abandonment and reforestation.

Projections of deforestation and reforestation with annual flux of carbon in billions of tons 1950-2100

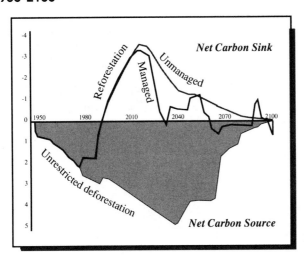

Present patterns of deforestation release from 0.4 to 3.5 billion tons of carbon to the atmosphere. Fossil fuels are responsible for another 5.4 billion tons annually. Reforestation of all available areas would provide a carbon sink for as much as 4 billion tons per year in the near term, but less over time. The value of the continuing carbon sink would depend on whether the new forest areas were managed (periodically harvested and replanted) or left as wilderness.

Source: Houghton, 1989.

An equal amount of land is available in Asia. A survey by the United Nations in 1985 determined that 831,000 square miles (2.15 million sq. km.) of land area in tropical southern Asian nations were not croplands, pastures, forests, or woodlands. A substantial part of this area—that portion that is not roads, railbeds, or settlements—consists of abandoned fields, such as the degraded grasslands that were once forest. Recent estimates place the portion of this wasteland available for reforestation at 580,000 sq. mi. (1.5 million sq. km.).

Potentially forestable land in Africa is more difficult to measure because much of Africa is not natural forest but grassland or desert. The unused lands that are in naturally forested areas amount to anywhere from 77,000 sq. mi. (200,000 sq. km.) to 580,000 sq. mi. (1.5 million sq. km.). If areas that were forested in ancient times could be reforested, the total potential for new forests in Africa would be 1.3 million sq. mi. (3.4 million sq. km.).

The total previously-forested land worldwide that is now available for reforestation is about 1.93 million sq. mi. (5 million sq. km.). Another 1.4 million sq. mi. (3.7 million sq. km.) could be obtained by shifting farm production from fallow land rotation to crop rotation and permaculture. The total amount of land potentially available for

Potential land use in India

While much of India is suitable for development of agriculture, parts of the subcontinent are so-called "wastelands" and grasslands that were once forested. These areas are suitable for large-scale reforestation.

Source: Joshi, 1989.

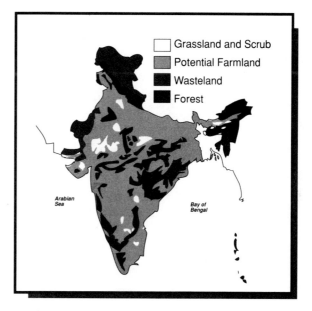

reforestation is therefore about 3.3 million sq. mi. (8.7 million sq. km.), 20 percent larger than the continent of Australia.

Today we are experiencing a steady rate of deforestation that will eliminate all tropical forests in the next 50 to 100 years. The alternative is to reduce cutting and start planting trees on a very large scale. If all of the available land were reforested in the next century, over 1 billion tons of carbon could be withdrawn from the atmosphere annually, approximately the amount that is being produced by forest-burning today. Much of the new forest could be harvested or grown in rotation, and as long as no more than about one-fifth of the forest products were burned or allowed to decompose, the world's carbon budget would balance.

INDIVIDUAL CHOICES

A new environmental ethic has to take root in the world, but social mores and reform cultures can be difficult to engineer. If we move too far too fast, we risk losing touch with human values and abilities, creating resentment and spawning backlash. The merits of reformist ideas have to be gauged by small experiments over long time frames. Social engineering takes time—a generation or more. And yet, there are ways to move things along.

21. REDUCE OVERCONSUMPTION.

Calculate what you, as an individual, consume. If you drive an automobile, figure that every 10,000 miles you drive at 25 miles per gallon produces 8,000 pounds of carbon dioxide, an amount which will be removed from the atmosphere only after roughly 145 tree-years (a 100 year old tree contains about 5,500 lbs. or 2,500 kg. of carbon). By increasing your mileage to 35 mpg, you produce only

Activities that generate one pound of carbon dioxide

Using a 100-watt lightbulb one evening
Driving a 20 mpg car one mile
Flying one mile in an airplane
Using one kilowatt-hour of electricity
Using 60 cubic feet of natural gas
Eating half a hamburger

A fast growing tree removes 55 lbs. of carbon from the air each year.

5,700 pounds of carbon dioxide, about 100 tree-years. The indirect costs—the carbon dioxide released to build and maintain the car, refine the gasoline, and bring those products and services to you, are about equal to the direct costs. If your car is air-conditioned, you are also contributing about 1 pound of CFCs to the atmosphere each year. If you are driving 10,000 miles and not planting between 200 and 300 trees or otherwise compensating for the carbon emissions, you are a net carbon source and are personally speeding global warming.

Flying one mile in an airplane generates, directly and indirectly, approximately one pound of carbon dioxide per passenger—about 60 tree-years from New York to Los Angeles.

Save the Earth
A checklist for First Worlders

Plant trees.
Take an energy audit of your home and workplace.
Fully insulate, caulk, and weatherstrip your home.
Turn down the thermostat in the winter, especially when not at home.
Install energy efficient windows, storm windows, and curtains.
Use compact fluorescent bulbs.
Turn out lights when not in a room.
Use fans instead of air conditioners.
Plant more trees.
Run your dishwasher only when full.
Use cold water instead of hot water whenever possible.
Presoak laundry, use less detergent, and sun-dry your clothes.
Take showers instead of baths.
Improve indoor natural lighting and airflow patterns.
Avoid products that contain CFCs, halons, and nonrecyclable plastics.
Shop for food at the farmers market or cooperative.
Get involved in tree planting projects.
Eliminate display lighting when your business is closed.
Drive car sparingly and maximize fuel efficiency.
Use public transportation, bicycles, and walking whenever possible.
Plant ornamentals with large green leaves.
Reduce meat consumption.
Start a vegetable garden.
Recycle.
Plant more trees.

The grazing land in a tropical forest required to be cleared to produce a quarter-pound hamburger is 55 sq. ft. (5.1 sq. m.). Five hundred pounds of carbon dioxide are emitted by clearing 55 sq. ft. of tropical forest. For each hamburger you consume, plant 10 trees.

Suppose the price you pay for electricity is 7 cents per kilowatt hour, which was the average rate charged for commercial and industrial uses in the United States in 1989. For every $1 worth of electricity you use, you produce 29 pounds of carbon dioxide directly and another 29 pounds indirectly—a little more than one tree-year for each dollar spent.

Heating with home oil for a year sends an average of 6.5 tons of carbon to the atmosphere. If you heat with home oil, figure on planting another 235 trees.

Calculate the amount of carbon you send skyward each year. The average American directly generates about 18.4 tons, or 670 tree-years. World output is about 4 tons per capita, or 145 tree-years. Try cutting back your production. And try planting more trees. The idea is not to reduce your pleasure in life, but to develop a more graceful lifestyle. A Chinese villager may not be able to install a solar water heater or a thermally-efficient window because he cannot afford to. How about you? Can you afford to?

Think of energy conservation in terms of "negawatts." A negawatt is one million watts of energy that could be saved by becoming more efficient and more thoughtful. Substituting 10 fluorescent lights for ten incandescent bulbs that normally burn 5 hours each night can save one negawatt per year—about the same as planting 75 trees. Ten thousand homes, each saving a negawatt, can retire a coal-burning electric plant, the carbon-saving equivalent of planting 750,000 new trees.

Viewing your level of personal consumption in terms of tree-years is important not only as a measure of greenhouse effect, but also as a reminder that humans may not be the most important species on the planet any more.

Epilogue
Final Call

When the 20th century began, neither human numbers nor human science had the capacity to greatly alter planetary balances. We were fragmented into dichotomies of government and private sectors, politics and economics, military and civilian concerns. But the problems we face at the start of the 21st century cannot be separated out into neat, well-defined categories. We have dissolved most of the distinctions. Attempts to meet the problems of tomorrow with the institutional forms of the past are destined to end in frustration. There are no issues which are just "carbon dioxide" or "population" or "crime" or "sea level" or "militarism" or "development." The issues are all multi-disciplinary, transnational, and global. They are not problems of physics and chemistry. They are political, economic, and cultural. They are problems of justice, fairness, and responsibility.

Some things about the human character are relatively predictable. Climate change will become a cultural priority when people consider it serious, certain, and likely to affect them personally. After we become able to visualize simple and relatively direct impacts which climate change will bring, we will begin to alter our cultural ethic about consumption and acquisition of personal wealth. If we are very fortunate, people in all parts of the world, across all strata of class and culture, will find a measure of personal nobility and community respect in undertaking the sacrifices that will be required, in learning to live with less. People will take greater satisfaction in the elegance of simplicity.

Yet it is more likely that many of the sacrifices that will have to be made will not be popular. It is likely that much of the developing world will blame those countries that are most industrialized and most wasteful for having created their problems, despite whatever corrective measures the industrial world now undertakes.

The 21 proposals in the last chapter of this book will be called radical by some critics. "Radical" means, literally, "return to the roots" (Latin: *radix*). Radical proposals are exactly what are needed. Our entire human civilization is rooted in nature. This book urges a graceful, gradual, and comfortable return to those roots. The alternative is to wait and to let nature impose an agenda that will be neither graceful, gradual, nor comfortable.

While governments and scientists disagree about the scope and timing of the effect, we stand at the edge of an onrushing catastrophe. We have to wake up to the danger. The consensus of the effect and the likelihood of its speed are compelling enough. We have only a very slim chance of avoiding a completely disastrous rise in temperature during the next century. Anything we do to try to slow the onset of the crisis is completely warranted. The case cannot be overstated. We cannot move slowly in the face of this peril. The magnitude of risk is too high for inaction, or delay.

We have the uniquely human capacity to change, to adapt to changes around us, and to make conscious decisions about what our lives will be like. Governments must change. Institutions must adapt. But, ultimately, it will come down to individuals. What are *you* going to do?

Part Three

Afterword

Peter Schweitzer
Executive Director, PLENTY USA

In 1974, a group of young, latter-day pioneers, three years into building an intentional, alternative, agrarian village on 1,750 acres in rural Tennessee, created a non-profit charity they called "PLENTY."

Without formal training or experience in disaster relief or international development, PLENTY's staff of volunteers began responding to calls for help. At first, natural catastrophes such as hurricanes, tornadoes, and floods within the borders of the United States got PLENTY's full attention. Emergency workers and construction teams rushed off to scenes of destruction to help pick up the pieces. When catastrophes occurred in Honduras and Mexico in 1975, PLENTY went international. When a powerful earthquake devastated Guatemala in 1976, PLENTY found itself suddenly immersed in a disaster of enormous, unbelievable proportions and in a world and culture about which it had little knowledge and even less understanding. Meeting the Mayans of Guatemala changed PLENTY. It began awakening into the real "Third World" of insidious poverty and the "Fourth World," where highly evolved ancient cultures are locked in a day-and-night struggle with the horrors of repressive, racial genocide and the rising encroachments of the dominant societies that surround them. Over a four-year period in Guatemala, 200 PLENTY volunteers saw firsthand the politics of malnutrition, the dynamics of infant mortality, and the intransigence of institutionalized poverty. They witnessed the preventable hazards that conspire to debilitate the poor: intestinal parasites, protein deficiency, lack of accessible potable water, absence of primary health care, shrinking farmlands, deforestation, joblessness, racism, transnational exploitation, and on and on. The volunteers began to understand that poverty is a tenacious hydra with multiple heads and many lives.

Over a fifteen-year history, PLENTY has launched volunteer projects in twelve countries on four continents: from running an orphanage in Bangladesh to constructing a woodworking shop in Dominica; from piping running water to villages and building a soy dairy in the Guatemala highlands to training health care workers in Lesotho; from operating a free, community ambulance service in the South Bronx to installing drip irrigation for Oglala Lakota farmers on the Pine Ridge Reservation in South Dakota.

PLENTY has worked to empower disadvantaged communities with the tools of self-sufficiency, tools like solar energy, primary health care and midwifery, diversified local economies, soybean food-processing, nutrition education, radio communications, and reforestation. One of PLENTY's highest priorities is the protection of indigenous cultures. The survival of these precious cultures is inextricably linked to the survival of the planet.

In 1979, attorney and PLENTY board member Albert Bates founded a PLENTY law program called the Natural Rights Center. The NRC's mandate was to instigate projects on behalf of innocent victims of ignorance, negligence, and exploitation. Exhibiting an uncanny ability to digest piles of technical and legal information at the speed of light and laboring relentlessly at all hours for little or no pay, Bates was soon tangling with some intimidating behemoths. He challenged the nuclear industry in a lawsuit to shut down the entire nuclear fuel cycle because, he argued, it is unethical to kill people in order to generate electricity, and routine releases of radiation do just that. He took the Defense Nuclear Agency to court on behalf of soldiers who had been exposed to radioactive fallout during the atomic tests of the 1950s. He fought the deepwell injection of toxic waste by irresponsible corporations. He helped stop the U.S. Army Corps of Engineers' largest wetland destruction program. He demanded and won rights for prisoners, native peoples, and endangered species.

Ten years ago, the Center's research began turning up the early warning signs of an impending climate crisis of world-wide dimensions. As time passed and very little was done, the idea for this book was born. Even today, in the face of mounting evidence that the continued survival of life on Earth is at stake, world leaders are wavering, scientists are haggling over the numbers, and industries are dragging their feet. If something is going to be done, then surely we the people are going to have to make it happen. We have to start somewhere, and a good place to start is to try and understand what is going on.

Climate in Crisis is the product of a two-year effort to draw together research from widely differing scientific disciplines, distill it into a coherent whole, and fashion it into something a reasonably curious person would find interesting and helpful. From his small, rustic office fronting a dirt road in the hills of Tennessee, Bates plugged his modem into the global electronic village and mustered a network of volunteer researchers, editors, and scientific fact-checkers. The result of their collective efforts is a concise, understated warning of impending global catastrophe.

Before I read *Climate in Crisis* I knew the problem was serious, but I didn't know on what orders of magnitude. After reading it, I got the picture. How serious? Well, it looks like if we don't do the job, we're history, and the cockroach may say, "Good riddance!" The book makes it abundantly clear: it's time to wake up, heed the early warnings, quit our self-destructive ways. It's not like we're helpless. We weren't born yesterday. As a species we need to act our age. If the cold war is ending, it's none too soon, because this job is going to take all of us, together, at once.

Selected References

The following documents were selected from among the more than 3,000 reference sources used in preparing this book. These materials are recommended for further reading.

Chapter Two: The Greenhouse Century

Abrahamson, D.E., ed., 1989. *The Challenge of Global Warming.* Washington D.C.: Island Press.

Bacastow, R.B., C.D. Keeling, and T.P. Whorf. 1985. Seasonal amplitude increase in atmospheric CO_2 concentration at Mauna Loa, Hawaii, 1959-1982. *Journal of Geophysical Research* 90:10529-10540.

Bach, W. 1983. *Our Threatened Climate.* Norwell, MA: Kluwer Acad. Publ.

Barnola, J.M., D. Raynaud, Y.S. Korotkevich, and C. Lorius. 1987. Vostok ice core provides 160,000-year record of atmospheric CO_2. *Nature* 329: 408-414.

Blake, D.R., and F.S. Rowland. 1988. Continuing worldwide increase in tropospheric methane, 1978-1987. *Science* 239:1129-1131.

Bolin B., E. Degens, S. Kempe, and P. Ketner, eds. 1984. *The Global Carbon Cycle. Scope 13.* Chichester: John Wiley and Sons.

Bolin, B., and R.B. Cook, eds. 1986. *The Major Biogeochemical Cycles and Their Interactions. Scope 21.* Chichester: John Wiley and Sons.

Bolin, B., B.R. Doos, J. Jager, and R.A. Warrick, eds. 1986. *The Greenhouse Effect, Climatic Change and Ecosystems.* Chichester: John Wiley and Sons.

Brown, L., et al., eds. 1989. *State of the World.* New York: W.W. Norton

Carbon Dioxide Assessment Committee. 1983. *Changing Climate.* Washington, D.C.: National Academy Press.

Clark, W. C. 1989. Managing planet Earth. *Scientific American* 261(3):46-54.

Council on Environmental Quality. 1981. *Global Energy Futures and the CO_2 Problem.* The Council on Environmental Quality, Washington, D.C.

Cowen, R. 1988. Wheels within wheels. *Mosaic* 19:3/4:60-69.

Craig H., et al. 1988. The Isotopic Composition of Methane in Polar Ice Cores *Science* 242:1535-9

Cromie, W. 1988. Grappling with coupled systems. *Mosaic* 19:3/4:31403

Detwiler, R.P., and C.A. Hall. 1988. Tropical forests and the global carbon cycle. *Science* 239:4247.

Dickinson, R.E., and R.J. Cicerone. 1986. Future global warming from atmospheric trace gases. *Nature* 319:109-115.

Edelson, E. 1988. Laying the Foundation. *Mosaic* 19:3/4:31147

Facklam, M. and H. Facklam. 1985. *Changes in the Wind: The Earth's Shifting Climate.* New York: Harbrace.

Fisher, A. 1988. One Model to Fit All. *Mosaic* 19:3/4:52-59

Ford, M. 1982. *The Changing Climate: Responses of the Natural Flora and Fauna* London: Allen Unwin.

Gallant, R. 1979. *Earth's Changing Climate.* Boston: Macmillan.

Goldsmith, E. and N. Hilyard 1988. *Earth Report: The Essential Guide in Global Ecological Issues.* Los Angeles: Price Stern.

Graedel, T. E., and P.J. Crutzen. 1989. The changing atmosphere. *Scientific American* 261 (3) : 58-68.

Gribbin, J. 1978. *Climatic Change.* Cambridge: Cambridge University Press.

Hansen, J., I. Fung, A. Lacis, D. Rind, S. Lebedeff, R. Ruedy, and G. Russell. 1988. Global climate changes as forecast by Goddard Institute of Space Studies Three-Dimensional Model. *Journal of Geophysical Research* 93:9341-9364.

Henderson-Sellers, A. and R. Blong, *The Greenhouse Effect: Living in a Warmer Australia*. Kensington: New South Wales Press.

Heppenheimer, T. 1988. The sum of its parts. *Mosaic* 19:3/4:38-51

Houghton, R. and G. Woodwell. 1989. Global Climatic Change. *Scientific American* 260:436-44.

Keeling, C.D., R. Bacastow, and T. Whorf. 1982. Measurement of the concentration of carbon dioxide at Mauna Loa Observatory, Hawaii. In W. Clark, ed., *Carbon Dioxide Review* pp. 377-384.

Kellogg, W.W., and Schware, R. 1981. *Climate Change and Society: Consequences of Increasing Atmospheric Carbon Dioxide*. Boulder, CO: Westview Press.

Kerr, R.1989. Hansen vs. the World on the Greenhouse Threat *Science* 244:1041-1043

Kohlmaier, G.H., H. Brohl, E.O. Sire, M. Plochl, and R. Revelle. 1987. Modelling stimulation of plants and ecosystem response to present levels of excess atmospheric CO_2. *Tellus* 39B:155-170.

Koppel, T. 1988. *News From Earth: The Koppel Report Show #3*, ABC News, New York: Journal Graphics.

Lachenbruch, A.H. and Marshall B.V. 1986. Changing Climate: Geothermal Evidence from Permafrost in the Alaska Arctic *Science* 234:689-696.

Lamb, H. 1972. *Climate: Past, Present, and Future*. New York: Methuen.

Londer R.1988. Learning the language of climate change *Mosaic* 19:3/4:102-112.

Manabe, 5., and Wetherald, R.T. 1975. The effects of doubling CO_2 concentrations on the climate of a general circulation model. *Journal of Atmospheric Sciences* 32:3-15.

Manne, A.S. 1989. CO_2 emission limits: an economic analysis for the USA. Stanford University.

Mintzer, I. 1987. *A Matter of Degrees: The Potential for Controlling the Greenhouse Effect*. World Resources Institute, Washington, D.C.

Moss, M. and S. Rahman 1986. *Climate and Man's Environment* Dubuque, IA: Kendall-Hunt.

Nordhaus, W.D. 1989. The economics of the Greenhouse Effect. Yale University.

Patrusky, B. 1988. Dirtying the Infrared Window. *Mosaic* 19:3/4:24-37

Passell, P. 1989. Cure for the Greenhouse Effect: the costs will be staggering. *New York Times* 139:48059:A1. Nov. 19.

Pearman, G., ed. 1989. *Greenhouse: Planning for Climate Change*. East Melbourne, Australia: CSIRO Publications.

Pittock, A. 1987. Report on reports: the carbon dioxide debate. *Environment* 29:25-30

Prinn, R.G., P. Simmonds, R. Rasmussen, R. Rosen, F. Alyea, C. Cardelino, A. Crawford, D. Cunnold, P. Fraser, and J. Lovelock. 1983. The Atmospheric Lifetime Experiment *Journal of Geophysical Research* 88:8353-8367.

Radian Corporation. 1987. Emissions and Cost Estimates for Globally Significant Anthropogenic Combustion Sources of NO_x, N_2O, CH_4, CO, and CO_2. EPA, Research Triangle Park.

Radok, U. 1987. *Towards Understanding Climate Change: The J. O. Fletcher Lectures on Problems and Prospects of Climate Analysis and Forecasting* Boulder, CO: Westview Press.

Ramanathan, V. 1988. The Greenhouse Theory of Climate Change: A Test by an Inadvertent Global Experiment. *Science* 240:293-299.

Ramanathan, V. and Dickinson, R.E. 1979. The role of stratospheric ozone in the zonal and seasonal radiative energy balance of the earth-troposphere system. *Journal of Atmospheric Sciences* 36:1084-1104.

Ramanathan, V., R.J. Cicerone, H.B. Singh, and J.T. Kiehl. 1985. Trace gas trends and their potential role in climate change. *Journal of Geophysical Research* 90:5557 5566.

Rasmussen, R., Khalil M.A.K. 1984. Atmospheric Methane in Recent and Ancient Atmospheres: Concentrations, Trends and Interhemispheric Gradients. *Journal of Geophysical Research* 89:11599-11605 .

Raynaud, D., J. Chappellaz, J.M. Barnola, Y.S. Korotkevich, and C. Lorius. 1988. Climatic and CH4 cycle implications of glacial-interglacial CH4 change in the Vostok ice core. *Nature* 333:655-657.

Roberts, W. and E. Friedman 1982. *Living With the Changed World Climate*. San Francisco: Univ. Press.

Sancton, T., et al., eds. 1989. Planet of the Year: What on Earth are We Doing? *Time* 1331:24-30. Jan. 2.

Schneider, S.H. 1989. *Global Warming: Are We Entering the Greenhouse Century?* San Francisco: Sierra Club Books.

Schneider, S.H. 1989. The changing climate. *Scientific American* 261 (3):70-79.

Schneider, S.H. 1989. The Greenhouse Effect: Science and Policy. *Science* 243:771-81.

Schneider, S.H. 1988. Doing something about the weather. *World Monitor* 1:3:28-37.

Schneider, S.H. 1988. The Greenhouse Effect and the U.S. summer of 1988: cause and effect or media event? *Climatic Change* 13:113-115.

Seidel, S, and D. Keyes. 1983. *Can We Delay a Greenhouse Warming?* Office of Policy and Resources Management, U.S. Environmental Protection Agency, Washington, D.C.

Topping, J. and A. Helm 1988. *Preparing for Climate Change*. Rockville MD: Govt. Insts.

Topping, J., ed. 1989. *Coping With Climate Change*. Washington D.C.: Climate Institute.

Trabalka, J.R., and D.E. Reichle, eds. 1986. *The Changing Carbon Cycle: A Global Analysis*. New York: Springer-Verlag.

Trabalka, J.R., ed. 1985. *Atmospheric Carbon Dioxide and the Global Carbon Cycle*. U.S. Department of Energy, Washington, D.C.

Turner, B.L. 1989. The human causes of global environmental change. In: DeFries , R. and T. Malone, eds., *Global Change and Our Common Future: Papers from a Forum* Washington D.C.: National Academy Press.

UNEP (United Nations Environment Programme). 1987. *The Greenhouse Gases*. UNEP, Nairobi.

Woodwell, G. 1988. Rapid Global Warming: Worse with Neglect. Testimony before Senate Committee on Energy and Natural Resources

Chapter Three: Runaway!

Asimov, I. 1989. *Beginnings* New York: Walker and Company.

Bartusiak, M. 1988. Energizing the climate cycles. *Mosaic* 19:3/4:80-89

Bates, D. R. 1957. *The Earth and Its Atmosphere*. New York: Basic Books.

Broecker, W.S. 1987. Unpleasant surprises in the greenhouse? *Nature* 328:123-126.

Brooks, C. 1922. *The Evolution of Climate* New York: AMS Press.

Budyko, M. 1982. *The Earth's Climate: Past and Future* Washington: National Academy Press.

Budyko, M. I. and Golitsyn, G. S. 1988. *Global Climatic Catastrophes* New York: Springer-Verlag.

Byalko, A. 1987. *Our Planet: The Earth* Moscow: Mir.

Carter, E. and V. Seaquist. 1984. *Extreme Weather History and Climate Atlas* Tomball TX: Strode.

Charlson, R., J. Lovelock, M. Andreae, and S. Warren. 1987. Oceanic phytoplankton, atmospheric sulphur, cloud albedo and climate. *Nature* 326:655-661.

Emiliani, C. 1978. The cause of the ice ages. *Earth and Planetary Science Letters* 37:349-352.

Ephron, L. 1988. *The End* Berkeley: Celestial Arts.

Flohn, H. and R. Fantachi. 1984. *The Climate of Europe: Past, Present and Future* Norwell, MA: Kluwer Acad. Publ.

Frakes, L. 1980. *Climates Throughout Geologic Time* Amsterdam: Elsevier.

Frenzel, B. 1973. *Climatic Fluctuations of the Ice Age* Middletown NY: Univ. Press Books.

Genthon, C., J. M. Barnola, D. Raynaud, C. Lorius, J. Jouzel, N.l. Barkov, V.S. Korotkevich, and V.M. Kotlyakov. 1987. Vostok ice core: climatic response to CO_2 and orbital forcing changes over the last climatic cycle. *Nature* 329:414-419.

Gleick, J. 1987. *Chaos: Making a New Science* New York: Viking.

Hansen, J., I. Fung, A. Lacis, S. Lebedeff, D. Rind, R. Ruedy, G. Russel, and P. Stone. 1987. *Prediction of Near Term Climate Evolution: What Can We Tell Decision*

Makers Now? First North American Conference on Preparing for Climate Change, Washington, D.C.

Hoffman, J.S., D. Keyes, and J.G. Titus. 1983. *Projecting future sea level rise.* Washington, D.C.: Government Printing Office.

Imbrie, J. and K.P. Imbrie, 1979. *Ice Ages: Solving the Mystery* Short Hills NJ: Enslow.

Jones, P.D., T.M.L. Wigley, C.K. Folland, D.E. Parker, J K. Angell, S. Lebedeff, and J.E. Hansen. 1988. Evidence for global warming in the past decade. *Nature* 332:790.

Kerr. R. 1988. La Niña's big chill replaces El Niño. *Science* 241:1037.1038.

La Brecque, M. 1988. A global chemical flux *Mosaic* 19:3/4:90-101.

Lamb, H. 1985. *Climatic History and the Future*. Princeton: Princeton University Press.

Lashof, D. 1989. The dynamic greenhouse: Feedback processes that may infuence future concentrations of atmospheric trace gases and climatic change. *Climatic Change* 14:213-242.

Legrand, M.R., RJ. Delmas, and RJ. Charlson. 1988. Climate forcing implications from Vostok ice core data. *Nature* 334:418-420.

Lorenz, E.N. 1968. Climatic determinism. *Meteorological Monographs* 30:1-30.

Lovelock, J.E. 1988. *The Ages of Gaia: A Biography of our living earth.* New York: W. W. Norton.

MacCracken, M.C., and F.M. Luther, eds. 1985. *Projecting the Climatic Effects of Increasing Carbon Dioxide.* U.S. Department of Energy, Washington, D.C.

Malone, T. 1973. *Weather and Climate Modification: Problems and Progress* Detroit: Gale Research.

Manabe, S., and R. Stouffer. 1980. Sensitivity of a global climate model to an increase of CO_2 concentration in the atmosphere. *Journal of Geophysical Research* 85:5529-5554.

Mitchell, J.F.B. 1988. Local effects of greenhouse gases. *Nature* 332:399-400.

Monastersky, R. 1989. Ancient ice reveals sudden climate shift *Science News* 135:364.

Murray, B., et al 1988. *Earthlike Planets: Surfaces of Mercury, Venus, Earth, Moon, Mars.* New York: W.H. Freeman.

Rampino, M., et al. 1987. *Climate: History, Periodicity, and Predictability* New York: Van Nos Reinhold.

Revelle, R. 1983. Probable future changes in sea level resulting from increased atmospheric carbon dioxide. *Changing Climate*. Washington, D.C.: National Academy Press.

Rifkin, J. 1988. The Greenhouse Doomsday Scenario *Washington Post* Jul. 31.

Rose, C. 1981. *Forecast of Disaster: Our Changing Weather* New York: Zebra Books.

Rose, M., et al. 1981. *Past Climate of Arroyo Hondo, New Mexico, Reconstructed from Tree Rings* Santa Fe: School of Amer. Research.

Sagan, C. and G. Mullen 1972. Earth and Mars: Evolution of atmosphere and surface temperatures. *Science* 177 52-56

Schneider, S.H. and R. Londer. 1984. *The Coevolution of Climate and Life* San Francisco: Sierra Club Books.

Siegenthaler, U., and T. Wenk. 1984. Rapid atmospheric CO_2 variations and ocean circulation. *Nature* 308-624-626.

Chapter Four: The Rising Tide

Barth, M.C. and J.G. Titus. 1984. *Greenhouse Effect and Sea Level Rise: A Challenge for This Generation.* New York: Van Nostrand Reinhold Company.

Broadus, J. 1989. Impacts of future sea level rise. In: DeFries, R. and T. Malone, eds., *Global Change and Our Common Future: Papers from a Forum* Washington D.C.: National Academy Press.

Bruun, P. 1962. Sea level rise as a cause of shore erosion. *Journal of Waterways and Harbors Division* (ASCE) 1:116-130.

Dean, R.G. et al 1987. *Responding to Changes in Sea Level.* Washington, D.C.: National Academy Press.

Emmanuel, K.A. 1988. The dependence of hurricane intensity on climate. *Nature* 326:483-485.

Gornitz, V.S., 5. Lebedeff, and J. Hansen. 1982. Global sea level trend in the past century. *Science* 215:1611-1614.

Jacobson, J. 1989. Swept Away. *World•Watch* 20-6.

Kyper, T., and R. Sorenson. 1985. Potential impacts of sea level rise on the beach and coastal structures at Sea Bright, New Jersey. In: O.T. Magson, ed. *Coastal Zone '85*. New York: ASCE.

Lyle et al. 1987. *Sea Level Variations in the United States* Rockville, MD: National Ocean Service.

Manolo, E.B. 1977. *Agroclimatic survey of Bangladesh* Dhaka: Bangladesh Rice Research Inst./Intl. Rice Res. Inst.

Meier, M.F. et al. 1985. *Glaciers, Ice Sheets, and Sea Level*. Washington, DC: National Academy Press.

Roy, P. and J. Connell 1989. The greenhouse effect: where have all the islands gone? *Pacific Islands Monthly* 59:16:16-21.

Schneider, F.H. and R.S. Chen. 1980. Carbon dioxide forming and coastline flooding; physical factors and climactic impact. *Annual Review of Energy* 5:107-140.

Smith, J., and D. Tirpak, eds. 1989. *The Potential Effects of Global Climate Change on the United States*. Environmental Protection Agency, Report to Congress, Washington, D.C.

Titus, J.G. 1985. Sea level rise and the Maryland coast. In: EPA, *Potential impacts of sea level rise on the beach at Ocean City, Maryland*. Washington, D.C.: U.S. Environmental Protection Agency.

Titus, J.G. 1986. Greenhouse effect, sea level rise, and coastal zone management. *Coastal Zone Management Journal* 14:3:147-171.

Titus, J.G. 1987. The Greenhouse Effects, Rising Sea Level, and Society's Response. In: Devoy, R.J.N. *Sea Surface Studies* New York: Croom Helm.

Titus, J.G., T. Henderson, and J.M. Teal. 1984. Sea level rise and wetlands loss in the United States. *National Wetlands Newsletter* 6:4. Environmental Law Institute.

Wilms, R.P. 1988. The heat is on: The effects of global warming on coastal North Carolina, *AWRA Symposium on Coastal Water Resources*, May 23, 1988.

Wunch, C. 1988. The Global Environmental Protection Act of 1988, Testimony to the Senate Subcommittee on Environment and Public Works, Subcommittee on Hazardous Wastes and Toxic Substances, and Subcommittee on Environmental Protection (Sept. 14, 1988).

Chapter Five: Summer Heat

Barnett, T. et al. 1988 On the prediction of the El Niño of 1986-1987 *Science* 241:192-195.

Becker, R.J., and R.A. Wood. 1986. Heatwave. *Weatherwise* 39(4):195-6.

Breitenbeck, G.A. 1988. *EPA Workshop on Agriculture and Climate Change*. Washington, D.C.: U.S. Environmental Protection Agency.

Bridger, C.A., F.P. Ellis, and H.L. Taylor. 1976. Mortality in St. Louis, Missouri, during heat waves in 1936, 1953, 1954, 1955, and 1966. *Environmental Research* 12:38-48.

Bruske, E. 1988. 104 (phew!) degrees hottest in 52 years. *The Washington Post* 111(225):A1, A6. July 17.

Cromie, W. 1988. Grappling with coupled systems. *Mosaic* 19:3/4:31403

Crowe, R.B. 1985. *Effect of Carbon Dioxide Warming Scenarios on Total Winter Snowfall and Length of Winter Snow Season In Southern Ontario*. Atmospheric Environment Service, Canada.

Cure, J.D. 1985. Carbon dioxide doubling responses: a crop survey. In: Strain, B.R., and J.D. Cure, eds. *Direct Effects of Increasing Carbon Dioxide on Vegetation*. Washington, DC: U.S. Department of Energy.

Decker, W.L., V. Jones, and R. Achutuni. 1985. The impact of CO_2- induced climate change on U.S. agriculture. In: White, M.R., ed. *Characterization of Information Requirements for Studies of CO_2 Effects: Water Resources, Agriculture, Fisheries, Forests and Human Health*. Washington, DC: U.S. Department of Energy.

DeSilver, D. 1989. The cattle-drought connection. *Vegetarian Times* 143:42-9.

Dukes-Dobos, F., 1981. Hazards of heat exposure. *Scandinavian Journal of Work and Environmental Health* 73-83.

Ellis, F.P. 1972. Mortality from heat illness and heat-aggravated illness in the U.S. *Environmental Research* 5(1):1-58.

Frederick, K.D. 1986. *Scarce Water and Institutional Change*. Washington, D.C: Resources for the Future.

Frederick, K.D., and A.V. Kneese.1988. Reallocation by markets and prices. In: AAAS, *Climatic Variability, Climate Change, and U.S. Water Resources*. Washington, D.C.: Amer. Acad. for Advancement of Sciences.

Geraghty, J., D. Miller, F. Van Der Leeden, and F. Troise. 1973. *Water Atlas of the United States*. Water Information Center, Port Washington, N.Y.

Gibbons, D.C. 1986. *The Economic Value of Water*. Washington, D.C: Resources for the Future.

Glantz, M.H., and Ausubel, J.H. 1984. The Ogallala Aquifer and carbon dioxide: Comparison and convergence. *Environ. Conser.* 11(2):123-31.

Grant, L.D. 1988. Health effects issues associated with regional and global air pollution problems. World Conference on the Changing Atmosphere, Toronto.

Haile, D.G. 1988. Computer simulation of the effects of changes in weather patterns on vector-borne disease transmission. EPA/ORP, Washington D.C.

Hansen, J., I. Fung, A. Lacis, S. Lebedeff, D. Rind, R. Ruedy, G. Russell, P. Stone. 1988. Global climate changes as forecast by the Goddard Institute for Space Studies three-dimensional model. *Journal of Geophysical Research* 93(D8): 9341-9364.

Hirsch, J. 1988. As streets melt, cars are flummoxed by hummocks. *The New York Times* 137(47599):Bl, B5. August 16.

Hrezo, M.S., P.G. Bridgeman, and W.R. Walker. 1986. Integrating drought planning into water resources management. *Natural Resources Journal* 26:141-167.

Hurt, R.D. 1981. *The Dust Bowl*. Chicago: Nelson-Hall.

Johnson, D.E. 1988. *Livestock Emissions Estimates*. EPA Workshop on Agriculture and Climate Change, Washington, D.C.

Jones, T.S. et al. 1982. Morbidity and mortality associated with the July 1980 heat wave in St. Louis and Kansas City, MO. *Journal of the American Medical Association* 247:3327-3331.

Kalkstein, L.S. 1988. The impact of CO_2 and trace gas-induced climate change upon human mortality. Washington, D.C.: U.S. Environmental Protection Agency.

Kalkstein, L.S., R.E. Davis, J.A. Skindlov, and K.M. Valimont. 1986. The impact of human-induced climate warming upon human mortality: a New York case study. *Proceedings of the International Conference on Health and Environmental Effects of Ozone Modification and Climate Change*. Washington, D.C.: U.S. Environmental Protection Agency.

Keller, C.A., and R.P. Nugent. 1983. Seasonal patterns in perinatal mortality and preterm delivery. *American Journal of Epidemiology* 118:689-98.

Kimball, B.A. 1985. Adaptation of vegetation and management practices to a higher carbon dioxide world. In: Strain, B.R., and J.D. Cure, eds. *Direct Effects of Increasing Carbon Dioxide on Vegetation*. Washington, D.C.: Department of Energy.

Kutschenreuter, P.H. 1959. A study of the effect of weather on mortality. *New York Academy of Sciences* 22:126-138.

Lachenbruch, A. and B. Marshall. 1987. Probing the permafrost *Scientific American* 29:62.

Lewis, N. 1988. Two more heat records fall as summer of 1988 boils on. *The Washington Post* 111 (257):A1, A10, A11. August 18.

Linsley, R.K., and J.B. Franzini. 1979. *Water Resources Engineering*. New York: McGraw-Hill, Inc.

Liverman, D.M., W.H. Terjung, J.T. Hayes, and L.O. Mearns. 1986. Climatic change and grain corn yields in the North American Great Plains. *Climatic Change* 9:327-347.

Lockeretz, W. 1978. The lessons of the Dust Bowl. *American Scientist* 66:560-569.

Mahlman, J. 1989. Mathematical modeling of greenhouse warming: how much do we know? In: DeFries, R. and T. Malone, eds., *Global Change and Our Common Future: Papers from a Forum* Washington D.C.: National Academy Press.

Maurits la Rivière, J. W. 1989. Threats to the world's water. *Scientific American* 261 (3) : 80-94.

Mederski, H.J. 1983. Effects of heat and temperature stress on soybean plant growth and yield in humid temperature climates. In: Raper, C.D., and P.J. Kramer, eds. *Crop Reactions to Water and Temperature Stresses in Humid, Temperate Climates*. Boulder, CO: Westview Press, pp. 35-48.

Meisner, D.J. et al. 1987. An Assessment of the Effects of Climate Warming on Great Lakes Basin Fishes. *Journal of Great Lakes Research* 13(3):340-352.

Meo, M., ed. 1987. *Proceedings of Symposium on Climate Change in the Southern U.S.: Impacts and Present Policy Issues, Science and Public Prog*. Univ. of Oklahoma, Norman, OK.

Mount, G.A., and D.G. Haile. D.G. 1988. Computer simulation of population dynamics of the American dog tick, *Dermacentor variabilis* (Acarii ixodidae). *Journal of Medical Entomology*.

National Research Council. 1977. *Climate, Climatic Change, and Water Supply*. Washington, D.C.: National Academy Press.

National Research Council. 1988. *Estimating Probabilities of Extreme Floods*. Washington, DC: National Academy Press.

Schneider, K. 1988. Drought cutting U.S. grain crop 31% this year. *New York Times*. August 12. p. Al.

Newman, J.E. 1978. Drought impacts on American agricultural productivity. In: Rosenberg, N.J., ed. *North American Droughts*. Boulder, CO: Westview Press, pp. 43-63.

Oeschsli, F.W., and R.W. Buechley. 1970. Excess mortality associated with three Los Angeles September hot spells. *Environmental Research* 3:277-284.

Office of Technology Assessment. 1986. *Technology, Public Policy, and the Changing Structure of American Agriculture*. OTA, United States Congress, Washington, D.C.

Parry, M.L. 1978. *Climatic Change, Agriculture and Settlement*. Dawson: Folkstone.

Parry, M.L., T.R. Carter, N.T. Konijn, eds. 1988. *The Impact of Climatic Variations on Agriculture. Vol. 1. Assessments in Cool Temperate and Cold Regions*. Norwell, MA: Kluwer Acad. Publ.

Postel, S. 1986. *Altering the Earth's Chemistry: Assessing the Risks*. Worldwatch Paper 71. Washington, DC: Worldwatch Institute.

Rosenberg, N.J., ed. 1978. *North American Droughts*. Boulder, CO: Westview Press.

Rosenzweig, C. 1985. Potential CO_2-induced climate effects on North American wheat-producing regions. *Climatic Change* 7:367-389.

Ritchie, J.T., and S. Otter. 1985. Description and performance of CERES-Wheat: A user-oriented wheat yield model. In: Willis, W.O., ed. *ARS Wheat Yield Project*. USDA-ARS. ARS-38.

Shapley, H. 1987. *Climatic Changes: Evidence, Changes, and Effects* Ann Arbor: Books on Demand UMI.

Smith, J., and D. Tirpak, eds. 1989. *The Potential Effects of Global Climate Change on the United States*. Environmental Protection Agency, Report to Congress, Washington, D.C.

Sontaniemi, E., U. Vuopala, E. Huhta, and J. Takkunem. 1970. Effect of temperature on hospital admissions for myocardial infarction in a subartic area. *British Medical Journal* 4:150-1.

Stevens, W.K. 1989. Acid rain and fertilization linked to Greenhouse Effect *New York Times* (Oct. 3, 1989) Y:21.

The Washington Post. 1988. Warped rails checked in Amtrak wreck. 111(246):A5. August 7.

Trenberth, K., G. Branstator and P. Arkin 1988. Origins of the 1988 North American Drought *Science* 242:1640-5

U.S. EPA 1989. Ecological Effects of Global Climatic Change. Chapter 5 In: *EPA Global Climatic Change Program*. Washington, D.C.: U.S. Environmental Protection Agency.

Udall, J. 1989. Nature Under Glass *Sierra* 74:4:34-39

Udall, J. 1989. Turning Down the Heat *Sierra* 74:4:26-30

Waggoner, P. 1989. *Climate and Water*. AAAS report, Washington DC.

Warrick, R.A. 1984. The possible impacts on wheat production of a recurrence of the 1930s drought in the U.S. Great Plains. *Climatic Change* 6:5-26.

Warrick, R.A., and M.J. Bowden. 1981. The changing impacts of drought in the Great Plains. In: Lawson, M.P., and M.E. Baker, eds. *The Great Plains: Perspectives and Prospects*. Lincoln: University of Nebraska Press.

Wilford, J.N. 1988. Vast Persistent Pattern Spreading Heat Wave. *The New York Times*, Al; July 19.

Wilms, R.P. 1988. The heat is on: The effects of global warming on coastal North Carolina, *AWRA Symposium on Coastal Water Resources*, May 23, 1988.

Wilson, C.A., and J.F.B. Mitchell. 1987. Simulated climate and CO_2-induced climate change over Western Europe. *Climate Change* 10:11-42.

Wilson, J. 1982. *Ground Water* Philadelphia:Academy of Natural Sciences.

Wiseman, J., and J.D. Longstreth. 1988. The potential impact of climate change on patterns of infectious disease in the United States. Washington D.C. : EPA.

Chapter Six: The Sky is Falling!

Blum, D. 1988. Early nuclear bomb tests may have depleted ozone. *Sacramento Bee*. Dec.

Browne, M. 1989. Saving the Ozone Layer *Sacramento Bee*. D6. Mar. 21.

Consumers Union 1989. Ozone: Can We Repair the Sky? *Consumer Reports*. 322-6 May.

Crawford, M. 1988. EPA: Ozone Treaty Weak. *Science* 242:25

Crutzen, PJ, 1988. Tropospheric ozone: An overview. In Isaksen, I.S.A., ed. *Tropospheric Ozone*. Reidel, Dordrecht. 3-32.

Douglas, M. 1987. *Ozone Layer Peril* Denver: Pye Publ.

Farman, J., B. Gardiner, and J. Shanklin. 1985. Large losses of total ozone in Antarctica reveal seasonal ClO_x/NO_x interaction. *Nature* 315:207-210.

Gore, A. 1988. Unbearable Whiteness *New Republic* 12:3

Green, B. 1987. Policies on global warming and ozone depletion *Environment* 29:5.

Gregory, P. 1973. *Microbiology of the Atmosphere* Chichester: John Wiley and Sons.

Gribbin, J. 1988. *The Hole in the Sky* New York: Bantam Books.

Hammitt, J.K., F. Camm, P.S. Connell, W.E. Mooz, KA. Wolf, D. 1987. Chemicals that may deplete stratospheric ozone. *Nature* 330:712- 716.

Hammitt, J.K., K.A. Wolf, F. Camm, W.E. Mooz, T.H. Quin, and A. Bamezai. 1986. Product Uses and Market Trends for Potential Ozone-Depleting Substances, 1985-2000. EPA, Washington D.C. and Rand, Santa Monica.

Kerr, R. 1987. Winds, Pollutants Drive Ozone Hole *Science* 238:156-8.

Lovelock, J., R. Maggs and R. Rasmussen. 1972. Atmospheric dimethyl sulfide and the natural sulfur cycle *Nature* 237:452-453

MacCracken, M. 1987. The Chlorofluorocarbon Dilemma *Science* 238,4827:598

Makhijani, A., A. Makhijani, and A. Bickel. 1988. *Saving Our Skins: Technical Potential and Policies for the Elimination of Ozone-Depleting Chlorine Compounds*. Environmental Policy Institute and the Institute for Energy and Environmental Research, Takoma Park, MD.

McElroy, M. and R. Salawitch 1989. Changing Composition of the Global Stratosphere. *Science* 243:763-70

Patrusky, B. 1988. Dirtying the Infrared Window. *Mosaic* 19:3/4:24-37

Pool, R. 1988. The Elusive Replacements for CFCs. *Science* 242:666-8

Roberts, L. 1989. Does The Ozone Hole Threaten Antarctic Life? *Science* 244:288-9

Rowland, F.S. 1989. The role of halocarbons in stratospheric ozone depletion. In National Research Council, *Ozone Depletion, Greenhouse Gases and Climate change*. National Academy Press, Washington, D.C. 33-47.

Shea, C. 1988. *Protecting Life on Earth: Steps to Save the Ozone Layer*. Worldwatch Paper 87. Washington, D.C.: Worldwatch Institute.

Shea, C. 1989. Mending the Earth's Shield. *World•Watch* 6:27-32.

Solomon, S. 1989. The Earth's fragile ozone shield. In: DeFries, R. and T. Malone, eds., *Global Change and Our Common Future: Papers from a Forum* Washington D.C.: National Academy Press.

Titus, J., ed. 1986. *Effects of Changes in Stratospheric Ozone and Global Climate. Vol. 1: Overview.* Washington, D.C.: EPA.

Weisskopf, M. 1988. EPA Urges Halt in Use of CFCs *Washington Post* Sept. 27. A3.

Chapter Seven: Tumbling Down

Baker, R.S.B. 1949. *Green Glory: The Forests of the World* New York: A. A. Wyn.

Booth, W. 1988. Johnny Appleseed and the greenhouse. *Science* 242:19-20.

Booth, W. 1989. New Thinking on Old Growth *Science* 244:141-3

Brown, L., et al., eds. 1989. *State of the World* New York: W.W. Norton

Byrne, G. 1988 Let 100 Million Trees Bloom *Science* 242:371.

C. Palm, R. Houghton and J. Melillo. 1986. Atmospheric Carbon Dioxide from Deforestation in Southeast Asia. *Biotropica* 18(3):177-188.

Clawson, M. 1975. *Forests for Whom and for What? Resources for the Future.* Baltimore, MD: Johns Hopkins University Press.

Colinvaux, P. 1989. The Past and Future Amazon *Scientific American* 260:102-8

Cubbage, F.W., D.G. Hodges, and J.L. Regens. 1987. Economic implications of climate change impacts on forestry in the South. In: Meo, M., ed. *Proceedings of the Symposium on Climate Change in the Southern U.S.: Future Impacts and Present Policy Issues.* New Orleans, LA: U.S. Environmental Protection Agency.

Davis, M. and C. Zabinski. Rates of dispersal of North American trees: Implications for response to climatic warming. *Proceedings of the Conference on the Consequences of the Greenhouse Effect for Biological Diversity.* Washington, D.C.: National Academy of Sciences.

Detwiler, R., and C. Hall. 1988. Tropical forests and the global carbon cycle. *Science* 239:42-47

Dickinson, R.E., ed. 1987. *Geophysiology of Amazonia.* Chichester: John Wiley and Sons.

Dover, M., and L. Talbot. 1987. *To Feed the Earth: Agro-Ecology for Sustainable Development.* World Resources Institute, Washington, D.C.

Dudek, D. 1988. *Offsetting new CO_2 emissions.* Environmental Defense Fund, New York.

Ehrlich, P. and A.E. Ehrlich, 1983. *Extinction: The Causes and Consequences of the Disappearance of Species* New York: Random House.

Ehrlich, P., A.E. Ehrlich, and J.P. Holdren, 1970. *Ecoscience: Population, Resources, Environment* San Francisco: W.H. Freeman and Company.

Ellis, W. 1988. Brazil's Imperiled Rain Forest *National Geographic* 174:6:772-99.

Fearnside, P.M. 1985. Brazil's Amazon forest and the global carbon problem. *Interciencia* 10:179-186.

Fearnside, P.M. 1986. Spatial concentration of deforestation in the Brazilian Amazon. *Ambio* 15:74-81.

Fearnside, P.M. 1987. Causes of deforestation in the Brazilian Amazon. In R. Dickinson, ed. *The Geophysiology of Amazonia.* Chichester: John Wiley and Sons.

Ford, E. 1984. The dynamics of plantation growth. In G. Bowen and E. Nambiar, eds. *Nutrition in Plantation Forests.* London: Academic Press, 17-52.

Fosberg, M.A. 1988. Forest productivity and health in a changing atmospheric environment. In: Berger, A., ed. *NATO Symposium on Climate and Geosciences.* Dordrecht, The Netherlands: Reidel Publishing Company.

Gillis, M. 1989. Malaysia: Public policies and the tropical forest. In R. Repetto, and M. Gillis, eds. *Public Policy and the Misuse of Forest Resources*, Cambridge University Press, Cambridge, U.K.

Goldemberg, J., T. Johansson, A. Reddy, and R. Williams. 1987. *Energy for Sustainable Development.* World Resources Institute, Washington, D.C.

Gradwohl, J., and R. Greenberg. 1988. *Saving the Tropical Forests.* Smithsonian

Institution/Earthscan Books, Washington, D.C.

Houghton, R. 1987. Biotic changes consistent with the increased seasonal amplitude of atmospheric CO2 concentrations. *Journal of Geophysical Research* 92:4223-4230.

Houghton, R. 1988. The flux of CO_2 between atmosphere and land as result of deforestation and reforestation from 1850 to 2100. Woods Hole Research Center, Woods Hole, MA.

Houghton, R. 1988. Tropical lands available for reforestation. Woods Hole Research Center, Woods Hole, MA.

Houghton, R. 1989. The Future Role of Tropical Forests in Affecting the Carbon Dioxide Concentration of the Atmosphere. Woods Hole Research Center, Woods Hole, MA.

Houghton, R., R. Boone, J. Fruci, J. Hobbie, J. Melillo, C. Palm, B. Peterson, G. Shaver, G. Woodwell, B. Moore, D. Skole, and N. Myers. 1987. The flux of carbon from terrestrial ecosystems to the atmosphere in 1980 due to changes in land use: Geographic distribution of the global flux. *Tellus* 39B:122-39.

Houghton, R., R. Boone, J. Melillo, C. Palm, G. Woodwell, N. Myers, B. Moore, and D. Skole. 1985. Net flux of carbon dioxide from tropical forests in 1980. *Nature* 316:(6029)617-20.

Houghton, R., J. Hobbie, J. Melillo, B. Moore, B. Peterson, G. Shaver, and G. Woodwell. 1983. Changes in the carbon content of terrestrial biota and soils between 1860 and 1980: A net release of CO_2 to the atmosphere. *Ecological Monographs* 53:235-2

Jordan, W., R. Peters, and E. Allen. 1988. Ecological restoration as a strategy for conserving biological diversity. *Environmental Management* 12(1):55-72.

Kalpavriksh. 1985. The Narmada Valley Project—development or destruction? *The Ecologist* 15:(5-6).

Karliner, J. and B. Hall. 1987. *A Forgotten War* Greenpeace 12:4:11-6

Kiester, A.R., R. Lande, and D.W. Schemske. 1984. Models of coevolution and speciation in plants and their pollinators. *American Naturalist* 124(2):220-243.

Kuusela, K. 1987. Forest products—world situation. *Ambio* 16:80-85.

Lashof, D. 1988. Technical background on reforestation as a CO_2 sink. Memorandum from Environmental Protection Agency to World Resources Institute, Washington, D.C., August 5.

Lenssen, N. 1989. Reprieve for the Rain Forest? *World•Watch* 2,1:35-6

Lovejoy, T. 1979. Refugia, refuges and minimum critical size: problems in the conservation of the neotropical herpetofauna. In: Duellman, W., ed. *The South American Herpetofauna: Its Origin, Evolution and Dispersal*. University of Kansas Museum Natural Hist. Monograph 7:1-485.

MacArthur, R., and E. Wilson. 1967. *The Theory of Island Biogeography*. Princeton, NJ: Princeton University Press.

Maguire, A., and J. Brown, eds. 1986. *Bordering on Trouble: Resources and Politics in Latin America*. Bethesda, MD: Adler and Adler, Inc.,

Malingreau, J-P., and C. Tucker. 1988. Large-scale deforestation in the southeastern Amazon basin of Brazil. *Ambio* 17:(1).

Marland, G. 1988. The prospect of solving the CO_2 problem through global reforestation. Department of Energy, Office of Energy Research, Washington D.C. DOE/NBB-0082.

May, R.M. 1988. How many species are there on Earth? *Science* 241:4872;141-1449.

Meo, M., ed. 1987. *Proceedings of the Symposium on Climate Change in the Southern United States: Future Impacts and Present Policy Issues. May 28-29, 1987*. New Orleans, LA University of Oklahoma and U.S. Environmental Protection Agency.

Myers, N. 1984. *The Primary Source: Tropical Forests and Our Future* New York: W.W. Norton.

Myers, N. 1979. *The Sinking Ark*. New York: Pergamon Press.

Myers, N. 1986. Tropical Forests: Patterns of Depletion. In Prance, G. ed., *Tropical Rain Forests and the World Atmosphere*. Boulder, CO: Westview.

Myers, N. 1988. Tropical forests and climate.

Myers, N. 1989. The Heat is On *Greenpeace* 14,3:8-13

Palm, C., R. Houghton and J. Melillo. 1986. Atmospheric Carbon Dioxide from Deforestation in Southeast Asia. *Biotropica* 18(3):177-188.

Peters, R. 1989. Threats to biological diversity as the Earth warms. In: DeFries, R. and T. Malone, eds., *Global Change and Our Common Future: Papers from a Forum* Washington D.C.: National Academy Press.

Pickett, S.T.A., and P.S. White. 1985.*The Ecology of Natural Disturbance and Patch Dynamics*. New York: Academic Press, Inc. Harcourt Brace Jovanovich.

Plumwood, V. and R. Routley. 1982. World Rainforest Destruction: The Social Factors. *Ecologist* 12:1:4-22.

Prance, G.T. 1986. *Tropical Rain Forests and the World Atmosphere*. AAAS Selected Symposium No. 101. Boulder, CO: Westview Press.

Rainforest Action Network 1989. Rainforest Destruction and the Indigenous Peoples of Malaysia *World Rainforest Report* IV, 4:6

Rainforest Action Network. 1989. Taking aim on the IMF. *Action Alert 36*

Repetto, R. 1988. *The Forest for the Trees? Government Policies and the Misuse of Forest Resources*. World Resources Inst., Washington D.C.

Rich, B. 1989. The politics of tropical deforestation in Latin America: The role of the public international financial institutions. In S. Hecht and J. Nations, eds. *Social Dynamics of Deforestation: Processes and Alternatives*. Ithica: Cornell Univ. Press.

Roberts, L. 1989. Hard choices ahead on biodiversity. *Science* 241:1759-1761.

Roberts, L. 1989. How Fast Can Trees Migrate? *Science* 243:735-7.

Salati, E., R. Victoria, L. Martinelli, and J. Richey 1989. Deforestation and its role in possible changes in the Brazilian Amazon. In: DeFries, R. and T. Malone, eds., *Global Change and Our Common Future: Papers from a Forum* Washington D.C.: National Academy Press.

Sampson, R. 1988. ReLeaf for global warming. *American Forests* 9-14. Nov/Dec.

Sedjo, R., and A. Solomon. *Climate and forests. Workshop on Controlling and Adapting to Greenhouse Forcing*. Environmental Protection Agency and National Academy of Sciences, Washington, D.C.

Setzer, A., and M. Pereira. 1988. Amazon biomass burnings in 1987 and their tropospheric emissions. Brazilian Space Research Center, Sao Jose dos Campos, Brazil.

Shabecoff, P. 1986. World Lenders Facing Pressures From Ecologists *New York Times* D24. Oct. 30.

Shabecoff, P. 1988. U.S. Utility Planting 52 Million Trees *New York Times*. A9. Oct. 12.

Shands, W., and J. Hoffman. 1987. *The Greenhouse Effect, Climate change, and U.S. Forests*. The Conservation Foundation, Washington, D.C.

Simons, M. 1988. Vast Amazon Fires, Man-Made, Linked to Global Warming *New York Times* Aug.12.

Smith, J., and D. Tirpak, eds. 1989. *The Potential Effects of Global Climate Change on the United States*. Environmental Protection Agency, Report to Congress, Washington, D.C.

Spurr, S.H., and B.V. Barnes. 1980. *Forest Ecology, 3rd Ed.* New York: J. Wiley and Sons.

Vitousek, P. 1989. Terrestrial ecosystems. In: DeFries, R. and T. Malone, eds., *Global Change and Our Common Future: Papers from a Forum* Washington D.C.: National Academy Press.

Wilcox, B. 1982. Biosphere Reserves and the Preservation of Biological Diversity. Towards the Biosphere Reserve: Exploring Relationships Between Parks and Adjacent Lands. Parks Canada and U.S. National Park Service, Kalispell, Montana; June.

Wilson, E.O. 1988. *Biodiversity*. Washington, D.C.: National Academy Press.

Wilson, E.O. 1989. Threats to biodiversity. *Scientific American* 261 (3) : 108-116.

Winterbottom, R., and P. Hazelwood. 1987. Agroforestry and sustainable development: Making the connection. *Ambio* 16:(2-3).

Woodwell, G. 1988. Biotic Implications of Climate Change: Look to Forests and Soils. *Preparing for Climate Change, Proceedings of the First North American Conference on Preparing for Climate Change: A Cooperative Approach, October 27-29, 1987*. EPA, Washington,D.C.

Woodwell, G., J. Hobbie, R. Houghton, J. Melillo, B. Moore, B. Peterson, and G. Shaver. 1983. Global Deforestation: Contribution to Atmospheric Carbon Dioxide. *Science* 222:4628: 1081-1086.

Woodwell, G., R. Houghton and T. Stone 1986. Deforestation in the Brazilian Amazon Basin Measured by Satellite Imagery In Prance, G. ed., *Tropical Rain Forests and the World Atmosphere*. Boulder CO:Westview.

Woodwell, G., R. Houghton, T. Stone, and A. Park. 1986. Changes in the area of forests in Rondonia, Amazon Basin, measured by satellite imagery. In Trabalka, J.R. and D.E. Reichle, eds. *The Changing Carbon Cycle: A Global Analysis*. New York: Springer-Verlag.

World Bank. 1986. The World Bank's Operational Policy on Wildlands: Their Protection and Management in Economic Development. The World Bank, Washington, D.C.

Chapter Eight: Human Dimension

Asimov, I. 1980. *Earth: Our Crowded Spaceship* New York: Fawcett

Brown, L. 1988. *The Changing World Food Prospect: The Nineties and Beyond*. Worldwatch Paper 85. Washington, D.C.: Worldwatch Institute.

Brown, L., 1989. Feeding six billion *World•Watch* 2:5:32.

Brown, L. 1989. What does global change mean for society? In: DeFries , R. and T. Malone, eds., *Global Change and Our Common Future: Papers from a Forum* Washington D.C.: National Academy Press.

Brown, L., et al., eds. 1989. *State of the World* New York: W.W. Norton

Bryson, R. and T. Murray 1977. *Climates of Hunger: Mankind and the World's Changing Weather* Madison, WI: University of Wisconsin Press.

Commission on Population Growth and the American Future 1972. *Population and the American Future* New York: New American Library

Ehrlich, P.R., and A.H. Ehrlich. 1972. *Population, Resources, Environment*. San Francisco: W.H. Freeman and Company.

Garrett, W. 1988. Where Did We Come From? *National Geographic* 1744:434-7.

Holden, C. 1988. The Ecosystem and Human Behavior *Science* 242:663.

Ignatius, A. 1988. Beijing boom: China's birthrate rises again despite a policy of one-child families. *Wall Street Journal*. July 14, p. 1.

Jacobson, J. 1988. *Environmental Refugees: A Yardstick of Habitability*. Worldwatch Paper 86. Washington, D.C.: Worldwatch Institute.

Jacobson, J., 1989. Baby budget *World•Watch* 2:5:21.

Kelley, A.C. 1988. Economic consequences of population change in the Third World. *Journal of Economic Literature* 26:4:1685-1728.

Keyfitz, N. 1989. The growing human population. *Scientific American* 261 (3) : 118-126.

Leakey, R. and R. Lewin, 1977. *Origins: What New Discoveries Reveal About the Emergence of Our Species and Its Possible Future* New York: Dutton.

Lesly, P. 1979. *Selections from Managing the Human Climate* Chicago: Lesly Co.

Livingston, J. 1973. *One Cosmic Instant* Boston: Houghton Mifflin.

Masters, R.1989. Evolutionary Biology and Political Theory. AAAS Annual Meeting.

Matras, J. 1973. *Populations and Societies*. Englewood Cliffs,N.J.: Prentice-Hall.Inc.

Miller, R. 1989. Human dimensions of global environmental change. In: DeFries , R. and T. Malone, eds., *Global Change and Our Common Future: Papers from a Forum* Washington D.C.: National Academy Press.

Ophuls, W. 1977. *Ecology and the Politics of Scarcity* San Francisco: W.H. Freeman and Co.

Postel, S. 1987. Life, the great chemistry experiment *Natural History* 96:41-8

Putman, J. 1988. The Search for Modern Humans *National Geographic* 1744:438-77

Rotty, R.M., and C.D. Masters. 1985. Carbon dioxide from fossil fuel combustion: Trends, resources, and technological implications. In Trabalka, J.R., ed. *Atmospheric Carbon Dioxide and the Global Carbon Cycle*. DOE, Washington, D.C. 63-80.

Rushton. J. 1989. Evolutionary Biology and Heritable Traits. AAAS Annual Meeting.

Stevens, W. 1989. Acid rain and fertilization linked to greenhouse effect. *New York Times* Oct 3. A21.

Singer, S. 1971. *Is There An Optimum Level of Population?* Population Council. New York: McGraw-Hill

Snyder, G. 1981. Ink and Charcoal *CoEvolution Quarterly* 32:48-50.

Wagner, Richard H. 1971. *Environment and Man.* New York: W.W. Norton

Weiss, D. 1987. Maximizing Diets in Food Deficit Countries Using a Computer Model *Horizons* 64:21-7

World Conference on Environment. 1988. *The Changing Atmosphere: Implications for Global Security.* World Conference Statement, Toronto.

Zachariah, K.C., and M.T. Vu. 1988. *World Population Projections* 1987-88 Edition. Baltimore, MD: Johns Hopkins University Press.

Chapter Nine: A Shift in Emphasis

Adams, J. 1974. *Conceptual Blockbusting* San Francisco: W. H. Freeman.

Adams, R. M. 1987. Evidence that the world's climate may be changing irreversibly has not yet resulted in any coordinated response. *Smithsonian* 18:12.

Bateson, G. 1979. Mind and Nature: A Necessary Unity New York: E.P. Dutton.

Foundation on Economic Trends, 1988. The Global Greenhouse Network Information Packet.

McKibben, B. 1989. *The End of Nature* New York: Random House.

Nash, R. 1983. *Wilderness and the American Mind 3d ed.* New Haven: Yale University Press

Russell, B. 1945. A History of Western Philosophy and its Connections with Political and Social Circumstances from Earliest Times to the Present Day New York: Simon and Schuster.

Chapter Ten: Deep Ecology

Abbey, E. 1968. *Desert Solitaire. A Season in the Wilderness* New York: McGraw-Hill

Adler, M.J., 1985. *Ten Philosophical Mistakes* New York: Macmillan.

Ames, R.T. 1986. Taoism and the Nature of Nature *Environmental Ethics* 8:4:317-350.

Anglemyer, M. 1980. A Search for Environmental Ethics: An Initial Bibliography Washington, D.C.: Smithsonian Institution.

Attfield, R. 1983. *The Ethics of Environmental Concern* New York: Columbia University Press.

Barney, G.O., et al., eds. 1981. *The Global 2000 Report to the President* New York: Penguin.

Bates, A.K., 1988. The Karma of Kerma: Nuclear Waste and Natural Rights. *Journal of Environmental Law and Litigation* 3:9-44.

Bell, G. ed. 1976. *Strategies for Human Settlements: Habitat and Environment* Honolulu: University Press of Hawaii.

Berg, P. 1976. Amble towards Continent Congress. Planet Drum #4, reprinted in *Raise the Stakes* 10:12 (1984).

Berg, P. 1981. Devolving beyond global monoculture *CoEvolution Quarterly* 24:24-28.

Berg, P. 1986. Growing a life-place politics. *Raise the Stakes* 11:16.

Berg, P. ed., 1978. *Reinhabiting a Separate Country: A Bioregional Anthology of Northern California* San Francisco: Planet Drum.

Bergon, F. ed., 1980. *The Wilderness Reader* New York: New American Library.

Berman, M. 1981. *The Reenchantment of the World* Ithaca: Cornell University Press.

Berry, T. 1982. *Teilhard in the Ecological Age* Chambersburg, Pa.: Anima Books.

Berry, W. 1977. *The Unsettling of America* San Francisco: Sierra Club Books.

Bodian, S. 1982. Simple in Means, Rich in Ends: A Conversation with Arne Naess. *Ten Directions* (Institute for Transcultural Studies, Zen Center of Los Angeles) Summer/Fall.

Bookchin, M. 1980. Open Letter to the Ecology Movement. *Rain* April.

Bookchin, M. 1981. The concept of social ecology. *CoEvolution Quarterly* 32:14.

Bookchin, M. 1982. *The Ecology of Freedom* Palo Alto: Cheshire.

Brooks, P. 1980. *Speaking for Nature* Boston: Houghton Mifflin.

Brown, D.A. 1987. Ethics, Science, and Environmental Regulation. *Environmental Ethics* 9:4:331-350.

Cahn, R. 1978. *Footprints on the Planet: A Search for an Environmental Ethic* New York: Universe

Cairns, J., ed., 1981. *The Recovery Process of Damaged Ecosystems* Ann Arbor, Mi.: Ann Arbor Science.

Callicott, J.B. 1989 American Indian Land Wisdom? Sorting out the Issues. *J. Forest History* 33-1:35-42.

Callicott, J.B. 1980. Animal Liberation. *Environmental Ethics* 2:4

Callicott, J.B. 1982. Traditional American Indian and Western European Attitudes Toward Nature: An Overview. *Environmental Ethics* 4:293-318.

Callicott, J.B. 1986 The Metaphysical Implications of Ecology. *Environmental Ethics* 8:4:301-316.

Capra, F. 1982. *The Turning Point: Science, Society and the Rising Culture* New York: Simon and Schuster.

Capra, F. and C. Spretnak, 1984. *Green Politics* New York: E. P. Dutton.

Carmody, J. 1983. *Ecology and Religion: Toward a New Christian Theology of Nature* New York: Paulist Press.

Catton, W. 1980. *Overshoot: The Ecological Basis of Revolutionary Change* Urbana: University of Illinois Press.

Catton, W. and R. Dunlap, 1980. New Ecological Paradigm for Post-Exuberant Sociology. *American Behavioral Scientist* 24:15-48.

Cheney, J. 1989. Postmodern Environmental Ethics: Ethics as Bioregional Narrative. *Environmental Ethics* 11:2:117-134.

Cheng, C. 1986. On the Environmental Ethics of the Tao and the Ch'i. *Environmental Ethics* 8:4:351-370.

Clastres, P. 1977. *Society Against the State. The Leader as Servant and the Humane Uses of Power Among the Indians of the Americas* New York: Urizen Books.

Cobb, E. 1977. *The Ecology of Imagination in Childhood* New York: Columbia University Press

Cohen, M. 1983. *Prejudice Against Nature. A Guidebook for the Liberation of Self and Planet* Washington D.C.: National Audubon Society Expedition Institute.

Conviser, R. 1984. Toward an Agriculture of Context. *Environmental Ethics* 6:1:71-86.

Cox, H. 1966. *The Secular City: Secularization and Urbanization in Theoretical Perspective.* New York: Macmillan.

Deloria, V. 1973. *God is Red* New York: Grossett.

Deutsch, E. 1986. A Metaphysical Grounding for Nature Reverence. *Environmental Ethics* 8:4:293-300.

Devall, B. 1984. The Edge: The Ecology Movement in Australia. *Ecophilosophy Newsletter* 4.

Devall, B. and G. Sessions 1985. *Deep Ecology* Salt Lake City: Peregrine Smith.

Diamond, S. 1974. *In Search of the Primitive. A Critique of Civilization* New Brunswick, N.J.: Transaction.

Dinnerstein, D. 1976. *The Mermaid and the Minotaur* New York: Harper and Row.

Dodge, J. 1981. Living By Life. *CoEvolution Quarterly* 32:6-12

Dodson-Gray, E. 1982. *Patriarchy as a Conceptual Trap* Wellesley Ma.: Roundtable Press.

Dombrowski, D. 1984. *The Philosophy of Vegetarianism* Amherst:Univ. of Mass. Press.

Dubos, R. 1980. *The Wooing of The Earth* New York: Charles Scribner's Sons.

Eckholm, E. 1978. Wild Species vs. Man: The Losing Struggle for Survival *Living Wilderness* 42:11-22.

Edey, M. 1989. On the concept of nature and the natural path. *Annals of Earth* 7:1:20-22.

Egler, F. 1970. *The Way of Science. Toward a Philosophy of Ecology for the Layman* New York: Hafner.

Ehrenfeld, D. 1978. *The Arrogance of Humanism* New York: Oxford.

Ehrlich, P. 1974. CoEvolution and the biology of communities. *CoEvolution Quarterly*.

Eiseley, L. 1970. *The Invisible Pyramid: A Naturalist Analyzes the Rocket Century* New York: Chas.Scribner's.

Elder, F. 1970. *Crisis in Eden: A Religious Study of Man and the Environment* Boston: Abingdon Press.

Elgin, D. 1981. *Voluntary Simplicity: Toward A Way of Life That Is Outwardly Simple, Inwardly Rich* New York: Morrow.

Erikson, E. 1950. *Childhood and Society* New York: W. W. Norton, 1950.

Everndon, N. 1978. Beyond Ecology. *North American Review* 263:16-20.

Ferguson, D. and N. Ferguson 1983. *Sacred Cows at the Public Trough* Bend, Or.: Maverick Publications.

Ferguson, M. 1980. *The Aquarian Conspiracy: Personal and Social Transformation in the 1980s* New York: St. Martin's.

Flader, S. 1974. *Thinking Like a Mountain: Aldo Leopold and the Evolution of an Ecological Attitude Toward Deer, Wolves, and Forests* Columbia: University of Missouri Press.

Folsom, S.L.E. 1980. Gary Snyder's Descent to Turtle Island: Searching for Fossil Love. *Western American Literature* 15:103-121.

Foreman, D. 1985. *Ecodefense* Tucson, AZ: Earth First! Books.

Forrester, J. 1971. *World Dynamics* Cambridge, Ma.: Wright-Allen.

Forstner L. and J. Todd, eds., 1971. *The Everlasting Universe: Readings on the Ecological Revolution* Lexington, Ma.: Heath and Co..

Fox, S. 1981. *John Muir and His Legacy* Boston: Little, Brown, and Co.

Fox, W. 1984. Deep Ecology: A New Philosophy of Our Time? *The Ecologist*, 14:5:194-200.

Fukuoka, M. 1978. *The One-Straw Revolution An Introduction to Natural Farming* Emmaus, Pa.: Rodale Press.

Gabor, D. 1964. *Inventing the Future* New York: Knopf, A.A.

Garreau, J. 1981. *The Nine Nations of North America* Boston: Houghton Mifflin.

Golley, F.B. 1986. Deep Ecology from the Perspective of Environmental Science *Environmental Ethics* 9:1:45-56.

Goodman, R. 1980. Taoism and Ecology. *Environmental Ethics* 2:37-80

Graham, F. 1970. *Since Silent Spring* Boston: Houghton Mifflin.

Guha, R. 1989. Radical American Environmental Thought *Environmental Ethics* 11:1:71-84.

Gunter, P. 1974. The Big Thicket: A Case Study in Attitudes Toward Environment, in Blackstone, W.T., ed., *Philosophy and Environmental Crisis* Athens: Univ. of Georgia Press.

Hardin, G. 1972. *Exploring New Ethics for Survival: The Voyage of Spaceship Beagle* New York: Viking.

Hardin, G. and J. Baden, eds., 1977. *Managing the Commons* San Francisco: Freeman.

Hargrove, E.C. 1980. Anglo-American Land Use Attitudes. *Environmental Ethics* 2:2:121-148

Hargrove, E.C. 1989. *Foundations of Environmental Ethics*. Englewood Cliffs: Prentice-Hall.

Haught, J. 1986. The Emergent Environment and the problem of Cosmic Purpose *Environmental Ethics* 8:2:139-150.

Hecht, S. and A. Cockburn 1989. Lands, Trees and Justice: Defenders of The Amazon *The Nation* 248:20:695-700.

Hughes, J.D. 1983. *American Indian Ecology* El Paso: Texas Western Press.

Humphrey, C.R. and F.R. Butell, 1983. *Environment, Energy and Society* Belmont, Ca.: Wadsworth.

Hunter, R. 1979. *Warriors of the Rainbow: A Chronicle of the Greenpeace Movement* New York: Holt, Rinehart and Winston.

Ip, P. 1983. Taoism and the Foundations of Environmental Ethics. *Environmental Ethics* 5.

Janik, D.I. 1981. Environmental Consciousness in Modern Literature. In Hughes, J.D. and R. C. Schultz, eds., 1981. *Ecological Consciousness* Washington, D.C.: University Press of America.

Kalupahana, D.J. 1986. Man and Nature *Environmental Ethics* 8:4:371-380.

Kellert, S.R. 1984. Assessing Wildlife and Environmental Values in Cost-benefit Analysis. *Journal of Environmental Management* 18:355-363.

Kellogg, W. and S.H. Schneider. 1974. Climate stabilization: for better or worse? *Science* 186:1163-1172.

Kennard, B. 1982. *Nothing Can Be Done Everything Is Possible* Andover, Ma.: Brick House.

Koopowitz, H. and H. Kay, 1983. *Plant Extinction: A Global Crisis* Washington, D.C.: Stone Wall Press.

Krieger, M.H. 1973. What's Wrong With Plastic Trees? *Science* 179:451.

Krutilla J.V. and A.C. Fisher, 1975. *The Economics of Natural Environments: Studies in the Valuation of Commodity and Amenity Resources* Baltimore: Johns Hopkins Press.

Kuhn, T. 1970. *The Structure of Scientific Revolutions* 2d ed. Chicago: University of Chicago Press.

LaChapelle, D. 1978. *Earth Wisdom* Los Angeles: Tudor Press.

LaFleur, W. 1978. Sattva: Enlightenment for Plants and Trees in Buddhism. *CoEvolution Quarterly* 19:47-52.

Lao Tzu, c. 600 B.C. *Tao Te Ching*, translated by Archie Bahn, New York: Frederic Ungar, 1958; translated by Gia-Fu Feng and Jane English, New York: Vintage, 1971; translated by R.L. Wing as The Tao of Power, Garden City NY: Doubleday, 1986.

Lasch, C. 1978. *The Culture of Narcissism: American Life in an Age of Diminishing Expectations* New York: W. W. Norton.

Lash, J., K. Gillman and D. Sheridan, 1984. *A Season of Spoils: The Story of the Reagan Administration's Attack on the Environment* New York: Pantheon Books.

Leiss, W. 1972. *The Domination of Nature* New York: G. Braziller.

Leopold, A. 1949. *A Sand County Almanac* New York: Oxford.

Livingston, J. 1981. *The Fallacy of Wildlife Conservation* Toronto: McClelland and Steward.

Lovelock, J. 1972. Gaia as seen through the atmosphere. *Atmospheric Environment* 6:579-580.

Lovelock, J. 1979. *Gaia: A New Look At Life On Earth* New York: Oxford.

Lovelock, J. and L. Margulis 1974. Atmospheric homeostasis by and for the biosphere: the Gaia hypothesis. *Tellus* 26:1-10.

Lovelock, J. and L. Margulis 1974. Homeostatic tendencies of the earth's atmosphere. *Origins of Life* 1:12-22.

Lowe, P. and J. Goyder, 1983. *Environmental Groups in Politics* London: George Allen.

Luoma, J. 1984. *Troubled Skies, Troubled Waters* New York: Viking.

Margulis, L., and J. Lovelock 1974. Biological modulation of the Earth's atmosphere. *Icarus* 21:471-489.

Margulis, L., and J. Lovelock 1975. The atmosphere as circulatory system of the biosphere—the Gaia hypothesis. *CoEvolution Quarterly*.

Martin, C. 1978. *Keepers of the Game* Berkeley: University of California Press.

Marx, L. 1970. American Institutions and Ecological Ideals. *Science* 170

Matthiessen, P. 1981. *Sand Rivers* New York: Viking Press.

Matthiessen, P. 1984. *Indian Country* New York: Viking.

McDonald, E. ed., 1936. *Phoenix* New York: Macmillan.

Meadows, D.H., et al., 1972. *The Limits to Growth* Washington, D.C.: Potomac Associates.

Merchant, C. 1980. *The Death of Nature: Women, Ecology and the Scientific Revolution* San Francisco: Harper and Row.

Meyer, P., 1986. Bioregions and econoregions. *Raise the Stakes* 11:8.

Milbrath, L. 1984. *Environmentalists* Albany: State University of New York Press.

Miller, G.T. 1983. *Living in the Environment* 3d ed. Belmont, Ca.: Wadsworth, 1983.

Milton, J. P. and M. T. Favor, 1971. *The Careless Technology* New York: Natural History Press.

Mirabehn. 1962. *The Thought of Mahatma Gandhi: A Digest* Ahmedabad: Navajivan.

Molesworth, C., 1983. *Gary Snyder's Vision* Columbia: University of Missouri Press.

Moline, J. 1986. Aldo Leopold and the Moral Community *Environmental Ethics* 8:2:99-120.

Naess, A. 1973. The Shallow and The Deep, Long-Range Ecology Movements: A Summary. *Inquiry* 16:95-100. Oslo.

Naess, A. 1974. *Gandhi and Group Conflict: An Exploration of Satyagraha, Theoretical Background* Norway: Universitetsforlaget.

Naess, A. 1979. Modesty and the Conquest of Mountains. in Michael Tobias, ed., *The Mountain Spirit* Woodstock, N.Y.: Overlook Press.

Naess, A. 1979. Self Realization in Mixed Communities of Humans, Bears Sheep and Wolves. *Inquiry* 22:231-242.

Naess, A. 1984. Intuition, Intrinsic Value and Deep Ecology. *The Ecologist* 14:5:201-204.

Nandy, A. 1983. The Pathology of Objectivity. *Ecologist* 1:3:202-207.

Needleman, J. 1975. *A Sense of the Cosmos: The Encounter of Ancient Wisdom and Modern Science* Garden City, N.Y.: Doubleday.

Nelson, R. 1983. *Make Prayers to the Raven. A Koyukon View of the Northern Forest* Chicago: University of Chicago Press.

North American Bioregional Congress 1987. *Proceedings* Forestville, CA: Hart Publishing.

Odell, R. 1980. *Environmental Awakening: The New Revolution to Protect the Earth* Cambridge, Ma.: Ballinger.

Olson, S. 1969. *Open Horizons* New York: Knopf.

Passmore, J. 1974. *Man's Responsibility for Nature* New York: Scribner's.

Passmore, J. 1975. *Attitudes Toward Nature*. Nature and Conduct ed. by Peters, R.S. New York: Macmillan.

Pearce, J. 1971. *The Crack in the Cosmic Egg* New York: Julian Press.

Pearce, J. 1977. *Magical Child* New York: E. P. Dutton.

Pinchot, G. 1947. *Breaking New Ground* New York: Harcourt, Brace and Co.

Pirages P., and P.R. Ehrlich, 1974. *Ark II: Social Response to Environmental Imperatives* San Francisco: Freeman.

Polyani, K. 1944. *The Great Transformation* Boston: Beacon.

Porritt, J. 1985. *Green: The Politics of Ecology Explained* New York: Basil Blackwell.

Pursell, C. 1973. *From Conservation to Ecology: The Development of Environmental Concern* New York: T.Y.L. Crowell.

Quarles, J. 1976. *Cleaning Up America: An Insider's View of the Environmental Protection Agency* Boston: Houghton Mifflin.

Raphael, R. 1981. *Tree Talk. The People and Politics of Timber* Covelo, Ca.: Island Press.

Rappaport, R., 1986. Restructuring the ecology of cities. *Raise the Stakes* 11:4.

Rawlings, D. 1980. Abbey's Essays: One Man's Quest for Solid Ground. *The Living Wilderness* 44-46. June.

Regan, T. 1981. The Nature and Possibility of an Environmental Ethic. *Environmental Ethics* 3:19-34.

Regan, T. 1983. *The Case for Animal Rights* New York: Random House.

Rifkin, J. 1981. *Entropy: A New World View* New York: Viking.

Rifkin, J. 1983. *Algeny* New York: Viking.

Rodman, J. 1977. The Liberation of Nature? *Inquiry* 20 Oslo.

Rodman, J. 1983. Four Forms of Ecological Consciousness Reconsidered. In T. Attig and D. Scherer, eds., *Ethics and the Environment* Englewood Cliffs, N.J.: Prentice-Hall.

Roszak, T. 1972. *Where the Wasteland Ends* New York: Anchor.

Roszak, T. 1978. *Person/Planet* Garden City, NY: Doubleday.

Roudey, R. and V. Roudey, 1978. Nuclear Energy and Obligations to the Future. *Inquiry* 21:2:133-179.

Routley, R. 1982. Roles and Limits of Paradigms in Environmental Thought and Action Research School of Social Sciences, Australian National University.

Rubin, C.T. 1989. Environmental Policy and Environmental Thought *Environmental Ethics* 11:1:27-52.

Russell, B., ed., 1945. *History of Western Philosophy* New York: Simon and Schuster, pp. 788-789, 827-828.

Sagoff, M. 1986. Process or Product? *Environmental Ethics* 8:2:121-138.

Santayana, G. 1926. The Genteel Tradition in American Philosophy. *Winds of Doctrine* New York: Scribner's.

Sax, J. 1980. *Mountains Without Handrails: Reflections on the National Parks* Ann Arbor, Mi.: University of Michigan Press.

Schnaiberg, A. 1980. *The Environment: From Surplus to Scarcity* New York: Oxford.

Schumacher, E.F. 1975. *Small Is Beautiful: Economics as if People Mattered* New York: Perennial Library, Harper and Row.

Seideal, S. and D. Keyes, et al., 1983. *Can We Delay A Greenhouse Warming?* Environmental Protection Agency.

Sessions, G. 1977. Spinoza and Jeffers on Man in Nature. *Inquiry* 20 Oslo.

Sessions, G. 1981. Shallow and Deep Ecology: A Review of the Philosophical Literature. In Hughes and Schultz, *Ecological Consciousness*.

Shepard, M. 1981. *Since Gandhi: India's Sarvodaya Movement* Weare, N.H.: Greenleaf Books.

Shepard, P. 1973. *The Tender Carnivore and the Sacred Game* New York: Scribner's

Shepard, P. 1977. A Sense of Place in American Culture. *North American Review* 262:22-32.

Shepard, P. 1978. *Thinking Animals: Animals and the Development of Human Intelligence* New York: Viking.

Shepard, P., 1983. *Nature and Madness* San Francisco: Sierra Club Books.

Sibley, M. 1977. *Nature and Civilization: Some Implications for Politics* Ithica: F. E. Peacock.

Skolimowski, H. 1981. *Eco-Philosophy* New York: Marion Boyars.

Snyder, G. 1980. *The Real Work* New York: New Directions.

Snyder, G. 1983. Good, Wild, and Sacred. *CoEvolution* Quarterly 39:17.

Steiner, S. 1976. *The Vanishing White Man* New York: Harper and Row.

Teale, E.W. 1954. *The Wilderness World of John Muir* Boston: Houghton Mifflin.

Thomas, L. 1974. *The Lives of a Cell* New York: Viking.

Udall, S. 1963. *The Quiet Crisis* New York: Holt, Rinehart, and Winston.

Van Den Bosch, R. 1978. *The Pesticide Conspiracy* Garden City N.Y.: Doubleday.

Van Strum, C. 1983. *A Bitter Fog: Herbicides and Human Rights* San Francisco: Sierra Club.

Varner, G.E. 1987. Do Species Have Standing? *Environmental Ethics* 9:1:57-72.

Wasserman, H. 1983. *America Born and Reborn* New York: Collier.

Watts, A. 1970. *Nature, Man and Woman* New York: Vintage.

Watts, A. 1975. *Psychotherapy East and West* New York: Vintage.

Westbrook, A. and 0. Ratti 1970. *Aikido and the Dynamic Sphere* Rutland, Vt.: Charles E. Tuttle.

White, L. 1967. Historical Roots of Our Ecologic Crisis. *Science* 155

Wittbecker, A. 1986. Deep Anthropology. *Environmental Ethics* 8:3:261-270.

Zimmerman, M. 1983. Toward a Heideggerian Ethos for Radical Environmentalism. *Environmental Ethics* 5

Chapter Eleven: The New Agenda

Adler, M. 1978. Aristotle for Everybody: Difficult Thought Made Easy. New York: Macmillan.

Clark, W. C. 1989. Managing planet Earth. *Scientific American* 261(3):46-54.

Earth Island Institute. 1988. Peace, Justice, and the Environment in Central America. San Francisco: Earth Island Institute.

French, H., 1989. An Environmental Security Council? *World•Watch* 2:5:6.

Gore, A. 1989. The global environment: a national security issue. In: DeFries, R. and T. Malone, eds., *Global Change and Our Common Future: Papers from a Forum* Washington D.C.: National Academy Press.

Chapter Twelve: 21 Better Ideas

Bates, A.K., 1988. Technological Innovation in a Rural Intentional Community, 1971-1987, *Bulletin of Science, Technology and Society* 8:183-199.

Bates, A.K., ed., 1978. *Honicker v. Hendrie: A Lawsuit to End Atomic Power* Summertown, TN: Book Publishing Co.

Bates, A.K., ed., 1979. *Shutdown: Nuclear Power on Trial* Summertown, TN: Book Publishing Co.

Beardsley, T. 1987. Vapor Lock *Scientific American* 257:32

Beatty, J. 1989. The exhorbitant anachronism *The Atlantic* 263:6:40-53

Berg, P., B. Magilavy and S. Zuckerman, 1989. *A Green City Program for San Francisco Bay Area Cities and Towns* San Francisco: Planet Drum Books.

Bertell, R., 1986. *No Immediate Danger?: Prognosis for a Radioactive Earth*, Summertown,TN: Book Publishing Co.

Bleviss, D. 1988. *The New Oil Crisis and Fuel Economy Technologies: Preparing the Light Transportation Industry for the 1990s.* New York: Quorum Press.

Boardman, D., 1989. *A Physician's Guide to the Health Effects of Ionizing Radiation* Acton, MA: Center for Atomic Radiation Studies (in press).

Brody, H. 1987. Energy-Wise Buildings. *High Technology*. February.

Brundtland, G. H. 1989. How to secure our common future. *Scientific American* 261 (3):190.

Brundtland, G., 1989. *Global Change and Our Common Future* Sixth Annual Benjamin Franklin Lecture, Washington D.C.: Smithsonian Institution.

Canine, C. 1989. Home Energy *Garbage* 1:2:20-27.

Carlson, D. 1988. Low-Cost Power from Thin-Film Photovoltaics. *Solarex*, Newtown, PA.

Carlson, Freedman and Scott. 1979. A Strategy for a Non-Nuclear Future. *Environment* 20:6:6.

Cartoceti, A. 1984. The impact of Nd-Fe-B on the typical applications of permanent magnets, in Mitchel, I., ed., *Nd-Fe Permanent Magnets: Their Present and Future Applications*. London: Elsevier.

Caufield, C. 1989. *Multiple Exposures: Chronicles of the Radiation Age* New York: Harper and Row.

Cavanaugh, R. 1986. Least-Cost Planning Imperatives for Electric Utilities and Their Regulators. *Harvard Environmental Law Review*. 10:299-344.

Cavanaugh, R., and E. Hirst. 1987. The Nation's Conservation Capital. *Amicus Journal* 38-42. Summer.

Center for Defense Information, 1989. The global network of United States military bases.*The Defense Monitor* 18:2:1-8.

Center for Renewable Resources, 1982. *The Reagan Energy Plan: A Major Power Failure* Washington,D.C.: National Audubon Society.

Clark, W. C. 1989. Managing planet Earth. *Scientific American* 261(3):46-54.

Council on Environmental Quality, 1979. *The Good News About Energy*. Washington D.C.: Government Printing Office

Crandall R.W., H.K. Gruenspecht, T.E. Keeler, and L.B. Lave. 1986. *Regulating the Automobile*. The Brookings Institution, Washington, D.C.

Critical Mass Energy Project. 1988. *Turning Down the Heat* Washington D.C.: Public Citizen.

Crosson, P. R., and N. J. Rosenberg. 1989. Strategies for agriculture. *Scientific American* 261 (3) : 128-135.

Dankoff, W. 1989. Efficient lighting for the independently powered home *Home Power* 9:20

Darrow, K., K. Keller and R. Pam 1981. *Appropriate Technology Sourcebook Volume II* Stanford:Volunteers in Asia.

Deluchi, M.A., R.A. Johnston, and D. Sperling. 1987.*Transportation Fuels and the Greenhouse Effect*. University Wide Energy Research Group, University of California, Berkeley, CA.

Edwards, T. 1989. An Ecologically Safe Air-Conditioning and Heat Pump Technology. SAE Technical Paper Series.

Ehrlich P., G. Daily, A. Ehrlich, P. Matson and P. Vitousek, Global change and carrying capacity: implications for life on Earth. In: DeFries , R. and T. Malone, eds., *Global Change and Our Common Future: Papers from a Forum* Washington D.C.: National Academy Press.

Ehrlich, P. 1984. The Nuclear Winter: Discovering the Ecology of Nuclear War, *Amicus Journal* 20-37. Winter.

Flavin, C. 1989. *Slowing Global Warming: A Worldwide Strategy*. Worldwatch Paper 91. Washington D.C.: Worldwatch Institute.

Forster, H.J. 1983. Big Cars, Too, Can Be Light On Fuel. *Resources and Conservation*. 10:85-102.

Forster, M. 1986. Two Cheers for the Agreement. *Environmental Policy and Law* 16(5): 142-143.

Frosch, R. A., and N. E. Gallopoulos. 1989. Strategies for manufacturing. *Scientific American* 261 (3) : 144-152.

Geller, H. 1986. Energy and Economic Savings from National Appliance Efficiency Standards. ACEEE, Washington, D.C.

Geller, H.S. 1988. *Residential Equipment Efficiency: 1988 Update*. American Council for an Energy Efficient Economy, Washington. D.C.

Gibbons, J. H., P. D. Blair, and H. L. Gwin. 1989. Strategies for energy use. *Scientific American* 261 (3) : 136-143.

Gleick, J., 1987. Superconductor May Yield Strongest Magnets, *New York Times* A1. Mar. 18.

Gofman, J.W., 1981. *Radiation and Human Health* San Francisco: Sierra Club Books.

Goldemberg J., T. Johansson, A. Reddy, and R. Williams. 1988. *Energy for a Sustainable World.* New Delhi: Wiley-Eastern.

Goldemberg, J, T. Johansson, A. Reddy, and R. Williams. 1987. *Energy for Development.* World Resources Institute, Washington, D.C.

Gore, A. 1989. World Environmental Policy Act of 1989. 101st Cong. 1st Sess. S. 201.

Gray, C. 1983. U.S. light vehicles—some exciting news for the 1990's. *Resources and Conservation* 10:65-84.

Gray, J., W. Davis, and J. Harned. 1988. *Energy Supply and Use in Developing Countries.* New York: University Press.

Hackleman, M. 1988. The hybrid electric vehicle *Home Power* 8:5

Hackleman, M. 1989. The hybrid-configured electric vehicle *Home Power* 9:13

Hafemeister, D. et al., eds. 1985. *Energy Sources: Conservation and Renewables* New York: American Institute of Physics.

Hafner, E. 1979. Human Equivalent Power *Environment* 20,6:34-6.

Harvey, L.D. 1988. Potential role of electric power utilities in hydrogen economy.*Workshop on Global Climate Change: Emergency Energy Technologies for Electric Utilities.* American Institute of Architects, Washington D.C.

Hay, N. P. Wilkinson, and W. James. 1988. Global climate change and emerging energy technologies for electrical utilities: The role of natural gas. *Workshop on Global Climate Change: Current Electricity Supply Alternatives.* American Gas Assn., Washington D.C.

Heppenheimer, T. 1988. Forging the tools *Mosaic* 19:3/4:70-79.

Hirst, E., J. Clinton, H. Geller, and W. Kroner. 1986. *Energy Efficiency in Buildings: Progress and Promise.* American Council for an Energy Efficient Economy, Washington, D.C.

Hoffman, M. 1988. A Stand-Alone PV System Home. *Power* 7:5-7

Hostetler, J.A. 1980. *Amish Society*, 3rd Ed. Baltimore: Johns Hopkins Univ. Press.

Houghton, J. 1984. *The Global Climate.* Cambridge: Cambridge Univ. Press.

Houghton, R. 1988. The flux of CO_2 between atmosphere and land as result of deforestation and reforestation from 1850 to 2100. Woods Hole Research Center, Woods Hole, MA.

Houghton, R. 1988. Tropical lands available for reforestation. Woods Hole Research Center, Woods Hole, MA.

Houghton, R. 1989. The Future Role of Tropical Forests in Affecting the Carbon Dioxide Concentration of the Atmosphere. Woods Hole Research Center, Woods Hole, MA.

Houghton, R. and G. Woodwell. 1989. Global Climatic Change. *Scientific American* 260:436-44.

Hu, D., I. Oliker and F. Silaghy. 1984. Power Plant Modification for Cogeneration. *Fossil Plant Life Extension Workshop* Washington D.C: Electric Power Research Institute.

Joshi, G. 1989. Forest policy and tribal development *Cultural Survival Quarterly* 13(2):17.

Kamo, R. 1987. Adiabatic diesel-Engine Technology in Future Transportation. *Energy* 12:(10/11)1073-1080.

Kempton, W., and M. Neiman, eds. 1987. *Energy Efficiency: Perspectives on Individual Behavior.* ACEEE, Washington, D.C.

Keyfitz, N. 1989. The growing human population. *Scientific American* 261(3):118-126.

Komanoff, C. 1988. Increased Energy Efficiency, 1978-1986, *Science* 239:128.

Koshland, D. 1989. A Tax on Sin: The Six-Cylinder Car. *Science* 243,4889:281.

Lashof, D and D. Tirpak, eds. 1989 *Policy Options for Stabilizing Global Climate.* EPA Draft Report to Congress.

Lemons, J. 1984. Can the Carbon Dioxide Problem Be Resolved? *The Environmental Professional* 6:52-71.

Lovins A. and H. Lovins, 1981. *Energy/War: Breaking the Nuclear Link* San Francisco.Friends of the Earth.

Lovins, A., L. Lovins, F. Krause, and W. Bach. 1981. *Least-Cost Energy: Solving the CO_2 Problem.* Andover: Brick House.

Lovins, A. 1977. Cost-risk-benefit Assessments in Energy Policy, 45 *George Washington Law Review* S:911.

Lovins, A. 1989. Energy Options *Science* 243:12.

Lovins, A., and M. Shepard. 1988. Implementation Paper #1: *Financial Electric End-Use Efficiency.* Rocky Mountain Institute, Old Snowmass, CO.

Lovins, A., and R. Sardinsky. 1988. *The State of the Art: Lighting.* Rocky Mountain Institute, Old Snowmass, CO.

Machado, S., and R, Piltz. 1988. *Reducing the Rate of Global Warming: The State's Role*. Renew America, Washington, D.C.

MacKenzie, J., 1988. *Breathing Easier: Taking Action in Climate Change, Air Pollution, and Energy Insecurity*. World Resources Institute, Washington, D.C.

MacNeill, J. 1989. Strategies for sustainable economic development. *Scientific American* 261 (3) : 154-165.

Marinelli, J. 1989. Cars *Garbage* 1:2:28-37.

McGrory, M. 1989. Step on the Methanol *Washington Post* A2. May 4.

Mellor, J. 1988. Sustainable Agriculture in Developing Countries *Environment* 30,9:7-20, 36-40.

Meyers, S. 1988. *Transportation in the LDCs: A Major Area of Growth in World Oil Demand*. Lawrence Berkeley Laboratory, Berkeley, CA. March.

Mintzer, I.M. 1988. *Projecting Future Energy Demand in Industrialized Countries: An End-use Oriented Approach*. EPA, Washington, D.C.

Morrison, D. 1989. The build-down.*The Atlantic* 263:6:60-64.

National Energy Conservation Center. 1986. *Potential for Energy Efficiency Improvements in the Commercial and Industrial Sectors*. National Energy Conservation Center, Washington, D.C.

New England Energy Policy Council. 1987. *Power to Spare: A Plan for Increasing New England's Competitiveness Through Energy Efficiency*. Energy Policy Council, Boston, MA.

Nivola, P. 1986. *The Politics of Conservation*. Washington, D.C.: Brookings Institution.

Nuclear Information and Resource Service 1988. Living in a Greenhouse World. *Groundswell* 10:3/4:1-7

OTA (Office of Technology Assessment). 1982. *Increased Automobile Fuel Efficiency*. OTA, Washington, D.C.

Parsoons, J. 1988. The scourge of cows *Whole Earth Review* 58:40.

Pierce, D., E. Barbier, and A. Markandya, 1989. *Sustainable Development: Economics and Environment in the Third World* London: Edward Elgar Pub. Ltd.

Poole, R. 1988. Solar cells turn 30. *Science*. 241:900-901.

Rader, N., K. Boley, D. Borson, K. Bossong and S. Saleska 1989. *Power Surge: The Status of Near-Term Potential of Renewable Energy Technologies* Washington D.C.: Public Citizen.

Renner, M. 1988. *Rethinking the Role of the Automobile* Worldwatch Paper 84. Washington D.C.: Worldwatch Institute.

Renner, M., 1989. What's Sacrificed when we arm *World•Watch* 2:5:9.

Rifkin, J., ed., 1988. *The Greenhouse Crisis: A Citizen's Guide*. Washington D.C.: Greenhouse Crisis Foundation.

Rocky Mountain Institute. 1986. *Advanced Electricity-Saving Technologies and the South Texas Project*. Rocky Mountain Institute, Old Snowmass, CO.

Rosenfeld, A.H. 1988. Energy Conservation, Competition and National Security, *Strategic Planning and Energy Management Journal*, 8:1:5-30.

Rosenfeld, A.H., and H. Hafemeister, 1988. Energy-efficient Buildings, *Scientific American*, 258:4:78-85.

Ruckelshaus, W. D. 1989. Toward a sustainable world. *Scientific American* 261 (3) : 166-175.

Sagan, C. et al., 1984. *The Cold and the Dark: The World After Nuclear War* New York: Norton.

Sant, R., 1981. *Eight Great Energy Myths* Arlington: Energy Productivity Center.

Sathaye J., B. Atkinson, and S. Meyers. 1988. *Alternative Fuels Assessment: The International Experience*. Berkeley, CA: LBL.

Satin, M. 1988. To balance the budget, build a sustainable society. *New Options* 48:1-8.

Schaefer, P. 1988. *Energy and Our Earth* New York: Ground Zero Club.

Schechter, L. 1989. Basic Principles of Solar Architecture *Home Power* 11:34-35.

Schell, J., 1982. *The Fate of the Earth* New York: Alfred A. Knopf.

Schipper, L., et al. 1989. Energy Use and Lifestyle: A Matter of Time? *Annual Review of Energy*. Annual Reviews Inc., Palo Alto, CA.

Schipper, L., S. Meyers, and H. Kelly. 1985. *Coming in from the Cold: Energy-Wise Housing in Sweden*. Washington, D.C: Seven Locks Press.

Schneider, C. 1989. Global Warming Prevention Act of 1989. 101st Cong. 1st Sess. H.R. 1078.
SERI (Solar Energy Research Institute). 1981. *A New Prosperity: Building a Sustainable Energy Future. The SERI Solar/Conservation Study*. Andover, MA: Brickhouse.
Seymour, J. and H. Girardet. 1987. *Blueprint for a Green Planet* New York: Prentice Hall.
Shea, C. 1988 *Renewable Energy: Today's Contribution, Tomorrow's Promise*. Worldwatch Paper 81. Washington D.C.: Worldwatch Institute.
Shea, C. 1988. Building A Market for Recyclables. *World•Watch*, 1:3, 12-18.
Stobaugh and Yergin, 1979. *Energy Futures: A Report of the Harvard Business School Project on Energy* New York: Random House.
Stubbing, R. and R. Mendel. 1989. How to save $50 billion a year. *The Atlantic* 263:6:53-59.
Summers, H. 1989. A bankrupt military strategy. *The Atlantic* 263:6:33-40.

Taylor, V., 1979. *The Easy Path Energy Plan* Cambridge: Union of Concerned Scientists.
Udall, J. 1989. Domestic Calculations. *Sierra* 74:4:33-34.
von Hippel, F., and B. V. Levi. 1983. Automotive Fuel Efficiency: the Opportunity and Weakness of Existing Market Incentives. *Resources and Conservation* 10:103-124.
Wasserman, H., N. Solomon, R. Alvarez and E. Walters 1982. *Killing Our Own: The Disaster of America's Experience with Atomic Radiation*. New York: Dell.
Wilkinson, C. 1988. The Idea of Sovereignty: Native Peoples, Their Lands, and Their Dreams. *Native American Rights Fund Legal Review* 13:4:1-11.
World Comission on Environment and Development. 1988. *Our Common Future* Oxford: Oxford University Press.
WRI (World Resources Institute) and IIED (International Institute for Environment and Development), 1988. *World Resources 1988-89*. New York: Basic Books, Inc.

Epilogue: Final Call

Abramson, R. 1989. U.S.-led bloc averts 'greenhouse' pollution deadline. *Sacramento Bee* A15. Nov. 8.
Durning, A. 1989. Groundswell at the grass roots. *World•Watch* 2:6:16-23.
Passell, P. 1989. Cure for the Greenhouse Effect: the costs will be staggering. *New York Times* 139:48059:A1. Nov. 19.

Ingram, H., H. Cortner, and M. Landy, 1989. Agenda Setting and Public Policy Related to Global Climate Change. AAAS Annual Meeting.
Renner, M. 1989. Forging environmental alliances. *World•Watch* 2:6:8-15.

Glossary of Terms Used

Acid rain:

precipitation which is made acid by combination with atmospheric hydrates, sulfides, and nitrates; capable of reacting with metals to form salts.

Adiabatic:

functioning without gaining or losing heat.

Agribusiness:

large-scale, capital-intensive method of food production.

Albedo:

fraction of solar radiation reflected by the Earth's surface, mostly by ice and snow.

Anthropocentric:

human-centered; based on the interest and outlook of humans.

Aquifer:

an underground watercourse or reservoir.

Assimilation:

in sociology, when a minority or immigrant culture adopts the lifestyle of a majority or dominant culture.

Barrier islands:

islands which parallel a coastline and provide shelter for inland beaches, marshes, and waterways.

Beach nourishment:

transporting sand along coastlines to prevent storm erosion of beaches.

Bermuda High:

area of frequent high atmospheric pressure above the Atlantic Ocean southeast of Florida.

Bike paths:

roadways or traffic lanes designated for the exclusive use of bicyclists.

Biocentric:

life-centered; not based on the interest and outlook of humans to the exclusion of other species.

Biological diversity:

having a large variety of different lifeforms; a sign of health in ecological systems.

Biomass:

carbon or potential fuel content of living things.

Boat people:

ship-borne refugees from regions bordered by water; typically in very poor and desperate condition.

Boreal:

pertaining to the forest or tundra zones of northern temperate and arctic regions.

Butterfly effect:

in chaos theory, the sensitivity of resulting physical occurrences to slight differences in initial conditions.

Carbon cycle:

periodic exchange of atmospheric, oceanic, and terrestrial carbon content which corresponds to global shifts in temperature, precipitation, glaciation, and other geophysical effects.

Carbon dioxide (CO_2):

a molecule composed of one atom of carbon and two atoms of oxygen.

Carbon monoxide (CO):

a molecule composed of one atom of carbon and one atom of oxygen.

Carbon tax:

a fee or levy placed on industrial emissions of carbon to the atmosphere.

Carrying capacity:

maximum number of users that can be supported by a finite system.

Carryover stocks:

resources, typically food, which are available in limited supply from one year to the next because of relatively constant or world-averaged production.

CFCs (chlorofluorocarbons):

chemical compounds like Freon which are commonly used as refrigerants, solvents, and aerosol propellants.

Chaos:

in physics, a theoretical study of the general patterns in random events.

Circumpolar vortex:

a wall of air currents which forms periodically around the polar regions, primarily due to the rotation of the Earth.

Cogeneration:

production of useful energy from waste heat.

Continental shelf:

an area of ocean floor, such as off the east coast of North America, where relatively shallow waters extend several hundred miles out to sea.

Cretaceous-Pleistocene boundary:

an epoch of history several million years past, marked by the disappearance of dinosaurs and the appearance of early man.

Cultural metaprograms:

consciously or subconsciously instilled directives which conform individuals to the traditions of their society.

Debt for nature:

a method of reducing international development loans in exchange for guarantees that important biophysical resources, such as rainforests, will be preserved.

Deep ecology:

an environmental ethic which places greater importance upon biocentrism than anthropocentrism.

Deforestation:

removal or reduction of forested areas.

Desertification:

creation or enlargement of desert areas.

Dike:

an earthwork designed to keep low-lying regions from being flooded by rivers and seas.

DMS (dimethylsulfide):

a compound emitted by phytoplankton which is important to the rate of cloud formation, albedo, and precipitation.

DNA (deoxyribonucleic acid):

protein compound which carries the genetic code for cellular reproduction.

Drought:

an extended period without normal rainfall.

Dust Bowl:

an area where soils are reduced to shifting sand and dust by extended drought, such as occurred in the center of North America in the 1930s.

El Niño (ENSO):

a periodic shift in ocean and atmospheric currents in the Pacific region characterized by torrential rains in South America and drought in North America.

Energy crops:

crops grown for their value as fuels, such as jojoba or babassú palm.

Energy slave:

an amount of externally-derived energy equal to the total power level of human metabolism.

Entrainment:

in physics, the tendency for an object to follow an established pattern and to resist forces which tend to shift it away from that pattern.

Environmental refugees:

people who flee from one region to another to escape ecological catastrophe.

Estuaries:

inland bays, swamps, and marshes which receive saltwater from ocean tides.

Faint early sun paradox:

inconsistent warmth of the Earth during a period when the sun was much cooler than it is now, believed by Sagan and Mullen to be due to a "supergreenhouse" effect.

Fence-to-fence farming:

practice of farming all available land rather than letting portions lie fallow to restore natural soil fertility.

Fertility rate:

rate at which a population grows, generally measured in number of children per adult female.

Fluidized bed combustion:

a method of increasing the efficiency of coal burning steam generators by suspending limestone gravel in the combustion chamber.

Freshwater lenses:

underground cavities of fresh water formed by geophysical processes on coral and volcanic islands.

GDFL:

Geophysical Fluid Dynamics Laboratory model of climate change developed by Wetherald and Manabe.

Gene:

a part of a cell chromosome the contains the hereditary information for development and reproduction.

Genetic engineering:

artificial transposition of genes to create modified, man-made organisms and lifeforms.

Geothermal energy:

useful energy derived from the heat deep within the Earth.

GISS:

Goddard Institute for Space Studies model of climate change developed by James Hansen and co-workers.

Glaciers:

mountainous ice formations created in periods when the rate of snowfall regularly exceeds the rate of snowmelt.

Goldilocks Phenomenon:

mildness and moderation of Earth's climate when compared to the climates of Venus and Mars, primarily as a result of the Greenhouse Effect.

Greenhouse shield:

moderating effect of Earth's atmosphere in keeping or shedding heat from the sun.

Hole in the ozone:

a sharp reduction of the ozone content of the atmosphere first observed over Antarctica in the mid-1980s.

Homo sapiens:

the family of man.

Hydroelectric power:

useful electricity or mechanical energy produced by employing the weight and momentum of river currents.

Ice cores:

samples taken by drilling holes into glaciers, a method used to determine the composition of the atmosphere prior to the age of scientific measurement.

Inclination of the Earth's axis:

a tilt toward or away from the sun which alters global climate in cycles of 23,000 years.

Indigenous peoples:

cultures or lineages which have continuously occupied particular regions irrespective of "civilizing influences" or recent changes in political borders.

Infant mortality:

rate of death in childhood, a public health index commonly used to measure standard of living.

Interglacial:

period of warm climate between successive ice ages.

Isostatic rebound:

the "bounce back" of land to the elevation it had before being depressed by the weight of glaciers.

Jet stream:

a high-altitude steering current in the atmosphere.

Jomon transgression:

a period of rapid sea-level rise in the western Pacific rim, occurring some 15,000 to 6,000 years ago.

La Niña:

a periodic shift in ocean and atmospheric currents in the Pacific region characterized by increased precipitation in North America.

Little Ice Age:

a drop in average temperature in Europe from 1645 to 1715, causing French agriculture to collapse.

Luddism:

blind opposition to mechanization and industrialization, from Ned Ludd, the hero of the revolt of British textile workers in 1811 and 1816.

Magnetohydrodynamics:

production of electric current by the conduction of hot gases through a magnetic field.

Marshall Plan:

a plan created by U.S. Secretary of State George Marshall to assist the reconstruction of Europe after the Second World War.

Methane (CH_4):

molecule composed of one atom of carbon and four atoms of hydrogen.

Methanol:

a fuel alcohol created by the distillation of wood.

Microbe:

a minute lifeform incapable of detection without magnification.

Microhydro:

a very small-scale use of hydro-power.

Mode locking:

in physics, the tendency for an object to follow an established pattern and to resist forces which tend to shift it away from that pattern.

Monkeywrenching:

thwarting environmental or social destruction by small acts of vandalism that cause damage to industrial property without endangering life.

Natural rights:

those liberties which all sentient beings are heir to; without which they languish in despair and with which they become happy, prosperous, and creative.

Nuclear winter:

theory that relatively small nuclear weapon exchanges involving city targets could catastrophically alter Earth's climate, creating widespread famine.

OSU:

Oregon State University model of climate change.

Ozone:

molecule consisting of three atoms of oxygen, formed naturally by radiation from the sun striking the upper atmosphere.

Passive solar:

architectural or energy designs which take advantage of sunlight without employing machinery or external power sources.

Permaculture:

agriculture relying primarily on crops that are hardy perennials and require little maintenance, tillage, and replanting.

Photolysis:

separation of oxygen from water by bombardment with ultraviolet light.

Photon:

unit of sunlight, having characteristics of both waveforms and particles.

Photosynthesis:

exchange of carbon dioxide and water for oxygen, performed by green plants in the presence of sunlight.

Photovoltaics:

generation of an electric current directly from the influx of sunlight.

Phytomass:

carbon or potential fuel content of plants.

Plankton:

microscopic plants (phytoplankton) and animals (zooplankton) that float and drift in great numbers on the surface of bodies of water.

Plutonium:

a man-made element produced by the fissioning of uranium, which is exceedingly toxic and long-lived.

Precipitation:

rain, snow, and hail.

Radioactivity:

production of destructive subatomic particles and waves by the decay of unstable elements.

Reforestation:

replanting forests in areas where earlier forests were cleared.

Renewable sources of energy:

useful sources of heat and motive power which derive from the Earth's daily income from the sun rather than from nonrenewable supplies of fuel.

Runaway greenhouse reaction:

a possible scenario in which the greenhouse effect stops being cyclical and becomes self-augmenting, resulting in ecological catastrophe.

Salinity:

salt content of a substance.

Sea ice:

ocean-borne floes of ice created by breakup of glaciers, and surface ice which forms on the ocean's upper layer in very cold climates.

Smog:

harmful, malodorous climate condition formed by emissions of air pollution from automobiles and industries.

Southern Oscillation:

a periodic shift in ocean and atmospheric currents in the Pacific region characterized by torrential rains in South America and drought in North America.

Star Wars:

a program of military research and development for the purpose of detecting and destroying missiles in flight from weapons platforms stationed in Earth orbit.

Stealth bomber:

an experimental aircraft design intended to be invisible or impervious to radar detection.

Superconductivity:

use of special alloys to reduce resistance in order to create high magnetic and electrical fields and to reduce transmission losses.

Supergreenhouse:

a solution posed by Sagan and Mullen to the faint early sun paradox, postulates that Earth's early atmosphere had a much stronger greenhouse effect than at present.

Superinsulation:

dramatic reduction of heat loss by buildings through the use of new designs and materials.

Supermagnets:

semicrystalline alloys, typically of neodymium, which align magnetic fields to the crystal lattice, thereby achieving great strength per unit of mass.

Sustainable development:

progress which seeks to achieve a healthy standard of living entirely from renewable resource cycles.

Telecommuting:

communication between the home and business by means of electronic media, such as computers, radio, television satellite, and telephone, reducing or eliminating the need to travel to and from the workplace.

Third World:

nations which are aligned with neither the marketplace industrialized powers of the West (NATO), nor with the central economy industrialized powers of the East (Warsaw Pact and China).

Trace gases:

greenhouse gases which are in minute concentrations in the atmosphere but which play a significant role in atmospheric chemistry.

Tree respiration:

process by which trees absorb carbon dioxide and other gases from the air and produce oxygen.

Troposphere:

region of the atmosphere closest to the Earth's surface

Tundra:

treeless area between arctic ice cap and boreal forests, characterized by low-growing vegetation such as lichens, mosses, and shrubs.

Typhoon:

hurricane occurring in the western Pacific or China Sea.

Ultraviolet:

light radiation with wavelength from 40 to 400 nanometers, which can be harmful to human health.

Urban heat island:

effect of densely populated areas being warmer than surrounding countryside, sometimes used to discount global warming as a product of measurement errors.

Vegetarian:

living on a diet consisting primarily of vegetable protein rather than animal products.

Wan-xi-shoa:

a chinese slogan meaning "later-longer-fewer," used to encourage birth control.

West Antarctic Ice Sheet:

large mass of ice, three miles thick and comprising nearly three-quarters of the world's fresh water supply; held suspended from the ocean by a series of small buttressing islands.

Wetland:

swampy or marshy area between dryland region and ocean or watercourse, which serves as water purification system and biological sanctuary.

Wind turbine:

device which draws useful energy from wind currents.

Younger Dryas:

period of sudden cooling and reglaciation in Europe which occurred from 11,000 to 10,000 years before present.

Index

acid rain 15, 19, 38, 82, 174
Adhémar 29
adiabatic diesel engines 165
aerosol sprays 11, 16, 76, 78
Afghanistan 119, 180
african game preserves 138
agoutis 88
agribusiness 177
agricultural capacity 115
agricultural subsidies 174
agriculture 14, 18, 43, 58, 90-91, 110, 134, 145, 149, 176-179, 184
air conditioning 11, 72-73, 78, 85, 166
airplanes 156, 179, 187
albedo 8, 15, 79-80
alcohol 115, 165
algae 28, 105
alternate-fuel aircraft 165
aluminum 13, 26, 164
Amazon 8, 21, 89, 92, 94, 138, 142, 195
American Southwest 19-20
Ames Research Center 75
ammonia 26-27
Anasazi 20
Antarctica 10, 16, 47, 76-77
anthropocentric 132-133, 138
ants that are protected by plants 88
aquifers 47, 50, 55, 57, 67, 140
Arara 155
Arctic 16, 62-63, 128
Aristotle 131-132, 138
Arizona 21
Arkansas 63
Armaments 179
Arrhenius 10
Asia 28, 91
Assateague Island 49
assimilation 156
Atlantic current 42, 45, 64
Australia 63, 107, 112, 155, 162, 183, 185

Australopithecus 104
automatic machine gun 135
automobiles 8, 14, 17, 117, 119, 126, 128, 139, 142, 156, 161-165, 172, 186
average temperature worldwide 6
axial tilt 28-33, 82
Azores 42
babassú palm 101
Babylonia 117
Bangkok 50
Bangladesh 58-59, 115, 119, 152
Barnola 10
barrier islands 54-55
Bay of Bengal 58
beach nourishment 53
beaches 47, 53, 100
beech 97
beef 175, 183
Beirut 158
Bengal Delta 58
Bermuda High 63
bicycles 154, 165
biocentrist 134
biological diversity 96, 110, 137, 174
biomass 171
birch 97, 99
birds that are protected by hornets 88
birth control 148, 151
birthrate 117, 149
Bleviss 162
Block Island 49
Boardman 158
boat people 59
boreal forests 31
Bradenton Beach 50
Brazil 18, 59, 87-94, 118-119, 138, 141
brazil nut tree 87, 88
brazilian bees 87
Britain 182
Broecker 45
Brooks Range 62
Bruun Rule 53
Brzenzinski 180

building design 166
burning 13
butterflies with 8-inch wingspans 89
butterfly effect 36
Cairo 155
California 154
California State University Chico 166
Callendar 10
Canada 19, 97, 119
canals of Venice 43
cancer 4, 83, 89, 113, 158
Cape Cod 49
Cape Hatteras 42
Cape of Good Hope 106
car air conditioners 85, 135
carbon absorption 90
carbon cycle 29-34
carbon dioxide 8-18, 25-30, 33, 38-39, 43, 60, 68, 81, 85, 89, 97-98, 112, 117-120, 126, 141-145, 156-157, 162, 165, 170-173, 183, 186-189
carbon monoxide 9, 13, 127, 162
carbon tax 173
carrying capacity 108, 116, 117, 164
carryover stocks 115
Cartoceti 169
cassava 177
caterpillars that masquerade as snakes 89
cattle 11, 18, 65-70, 87, 92-95, 114, 116, 156, 183
CFCs 11, 16, 73, 76, 79, 84-86, 127, 187
chaos 35-38, 159
charcoal 90, 156
Charleston, S.C. 55
chemical composition of the North Atlantic 44
Chesapeake Bay 49, 55
Chicago 72
chickpeas 177
China 112-120, 143, 149-150, 171, 182

chlorination of city water 72
chlorine and ozone layer 78-80, 85
chlorocarbons 9, 13
chlorofluorocarbons 11, 16, 73, 76, 79, 84-86, 127, 187
Cinta Larga 155
circumpolar vortex 76
Clark 140, 176
climate policy 174
coal and oil 11
coast of Alaska 46
coastal areas 53
cogeneration 160, 171
Colorado 20, 66, 154
Columbia University 45
compulsory sterilization 150
continental shelf 46
contraceptives 149
cooling 10, 22, 26-27, 32, 43, 63
Cooper 127
corn 19, 64-68, 89, 91, 109-110, 113, 128, 164, 175
Cortez 106
cotton 68
Cretaceous-Pleistocene boundary 96
crickets 38
Croll 29
Cromie 64
cultural metaprograms 125
cultural priority 190
cyanobacteria 27, 105
cyclone 59
Dallas 72, 141
Dankoff 161
debt for nature 94-95
deep ecology 132-134
deep ocean sediments 31
defense budget of the United States 181
deforestation 11, 17-19, 30, 58, 90, 92-94, 100, 128, 142, 147, 156, 184-185
Delaware River 55
Delta Plan 48
deltas 50
demand for electricity 72

demand for food 152
Denmark 118, 171
Denver 72
deserts 19-20, 88, 95, 138, 140, 155, 182, 184
Dhaka 155
diamond distribution of wealth 136
dikes 48, 58, 72
dimethylsulfide 81-82
dinosaurs 96
DMS 81-82
DNA molecule 4
domestic animals 11, 177
Dorian tribes 20
drip irrigation 177
droughts 14, 19-23, 62-64, 67, 70, 111-115
dust bowl 67
earlier snowmelt 14
early atmosphere 26
Earth 26
earthquake 154
Eastern Europe 19
ecological communities 97
Economic Summit in Paris 145
educational systems 152
Egypt 59, 181
Ehrlich and Holdren 141
Eisenhower 180
El Niño 62, 64
El Niño of 1982-83 63
electric cars 154, 162
electrical generation 30, 73, 156, 170
elegance of simplicity 190
Emanuel 48
emergency food aid 175
energy 156
energy crops 171
energy efficiency 159, 172
energy slaves 172
ENSO 63
entrainment 37
Environmental Protection Agency 70, 144
environmental refugees 153
equator 19, 21, 42-46, 63-64

equatorial forests 18
erosion 47, 53-54, 58, 67
estuaries 55-56
ethanol 171
Europe 28, 42, 44-45, 57, 70-71, 107, 115-116, 143-144, 163, 174, 182
evaporation 20, 44-45
Eve 105
Everglades 49
extreme poverty 136
faint early sun paradox 32
family planning 93, 149, 150-152
Farman 10, 16, 76-77
fence-to-fence farming 109
fertility rate 118, 120, 151
fertilization of the oceans 30
fertilizer 11, 110-112, 142, 176
firewood 156
fissioning of uranium 158
Florida 47-57, 63, 68, 100, 141, 163
fluidized bed combustion 160
food riots 115
food security 174
food surpluses 174
forest burning 9
forest fires 14-15, 19, 30, 94
Fort Lauderdale 54
fossil fuels 17, 27, 90, 111, 140, 145, 147, 156, 160, 166, 171-172, 183-184
Fourier 10
France 118, 182
French Guiana 59
freon 43
freshwater lenses 59
fungi that are protected by ants 88
Galveston Bay 54-55
Gandhi 132, 136
gasoline mileage standards 162
Gavião 155
GDFL 70
genes 4, 111
genetic code 4
genetic engineering 110, 111

Geophysical Fluid Dynamics Laboratory 70
geothermal energy 171
Geraghty and Miller 65
Gibbons 160
GISS 70
glaciers 31, 44-46, 91
Gleick 35-36
GM Sunraycer 169
Goddard Institute for Space Studies 10, 21, 70, 77
Goldilocks Phenomenon 8, 39
Grasslands 65
Great Britain 145
Great Lakes 97
Great Plains 19, 22, 64-67, 70
Great Salt Lake 21
Greece 171
greenhouse gases 8-11, 14-15, 22-24, 30-32, 43, 86, 112, 126, 138, 141, 147, 165, 170-174, 183
greenhouse shield 8, 9, 15
Greenland 52, 64, 74, 128
Greenland Ice Sheet 52
Guatemala 95, 141, 181
Gulf of Mexico 45, 57, 63, 71
Gulf Stream 42, 44
Guyana 59
Haile 69
Halley Bay 76, 77, 143
Hansen 10, 20-24, 33, 70, 72
Hawaii 46, 141
hazardous waste landfills 55
health care 152-156, 179-182
Hemingway 127
hemlock 97
hierarchies of wealth and power 136
Hilton Head 49
Hiroshima 158
Hittites 20
hole in the ozone 10, 16, 76-86, 121
Homo sapiens 103-108, 121
hornets that are protected by trees 88
hottest year in history 5

Houghton 93
hourglass distribution of wealth 136
human mortality 14, 72-73
human powered vehicle 166
hunters and gatherers 106
hurricanes 14, 42, 48, 54-55
Huxley 132
Huygens 37
hydrocarbon 9
hydroelectric power 169
Hydrogen 26-27, 34, 72, 79, 164-165
ice 4, 8, 10, 14-15, 28-33, 44-47, 51, 60-65, 79-80, 82, 91, 97, 105-107
ice ages 6, 28-33, 44-47, 60, 64-65, 91, 97, 105-107
ice in atmosphere 8, 79-80
ice cores 10, 82
iceland 42, 46
Imbrie 33
inclination of the Earth's axis 28-33, 82
India 118, 171, 182, 184
Indian Ocean 59, 70
indigenous peoples 155
Indonesia 63, 93, 118, 151
Indonesian birth control program 151
infant mortality 140, 148-153
infrared heat from Earth 14-16, 39, 78, 85
insects 11, 15, 68-69, 89, 97
Inter-American Development Bank 18, 94
interdependence 88
interglacials 31-32, 45, 65
international aid 174
International Monetary Fund 18, 95
Intracoastal Waterway 54
Iran 180
Iraq 180
isostatic rebound 46
Israel 177
Jakarta 154
Jamestown 106

Japan 143, 149, 158, 172, 177
Jefferson 137
Jet Propulsion Laboratory 79
jet stream 19, 21, 63
Jomon transgression 51
Jordan 119
Joshi 185
Jupiter 26
Kansas 66
Kayfitz 151
Keeling 9-10
Kenya 119
Key West 52
Kiribati 59
La Niña 63
land's memory 61
Langway 13
Lao Tsu 133
Latin America 183
leaking refrigerators 78
Leopold 132
limits of technology 141
Little Ice Age 43
livestock 114, 177
loggers 92
London 55
Lorenz 35-37
Lorenz attractor 36
Los Angeles 72
Louis XIV 43
Louisiana 47, 49, 52, 56-57
Lovejoy 97
lowering the greenhouse ceiling 85
Lucy 104-105
Ludd 135
Luddism 135-136
Lung-shan 20
luxury of mobility 165
Lyme disease 69
Machita 51
Madagascar 101-102
Maginot 135
magnetohydrodynamics 160
Malaysia 93, 128
Maldives 59
Malthus 108
manatees 100

Manolo 58
marketplace adjustments 137
Mars 7-8, 26-27, 33-34, 39, 56-57, 154
Marshall Plan 144
Marshall Islands 59
marshes 50
Martha's Vineyard 49
Massachusetts Institute of Technology 35, 48
Mauna Loa 9-10, 143
Maunder Sunspot Minimum 43
Maxum 135
McKibben 127
meat-packing plants 92
Mediterranean Sea 71
Memphis 72
Mercury 26, 39, 74
Mesa Verde region 20
Mesopotamia 117
meteorological models 19
methane 8-17, 25-33, 38, 79, 112, 115, 126, 170, 177
methanol 171
Mexico 118-119
Mexico City 155
Miami Beach 54
Michigan State University 66
microbes 11, 112
microhydro 171
Micronesia 59
middle class 136
Midwest United States 23, 63-64, 109, 113
migrating 17, 98
militarism 134, 179-182
military personnel 179
military research 181
millet 177
Mintzer 11
Mississippi 19, 100
Mississippi River 22, 45, 52, 65, 72
mode locking 37-38
Molina 16, 78-80
monkeywrenchers 133
monsoons 14, 58, 70
Montezuma 106

Montreal 11, 84, 85, 158
Morris 73
Muir 132
Mullen 32-33
Mycenean empire 20
Myers 90
Naess 132
Nantucket 49
NASA 23, 62, 75, 77, 79, 154
Nashville 72
National Center for Atmospheric Research 23, 62
national security 159
NATO 182
natural rights 132-133, 137
Nebraska 66
Neptune 26
Netherlands 48, 58
Nevada 72
New Jersey 63
New Mexico 66
New Orleans 49, 52, 72
New York 72
New York Times 180
Newfoundland 42
Nigeria 93
Nile Basin 52, 59
nitric acid 79
nitrogen 9, 13, 26-27, 80, 89, 111, 116, 127
nitrogen oxide 9
nitrous oxide 8-11, 126
normal range of variation 22
North America 19-21, 28, 54, 63-67, 70-75, 97-99, 107-108, 112-119, 133, 141, 166, 170-174
North Atlantic 45
North Atlantic Drift 42
North Carolina 49
North Equatorial Current 42
Northeast Trades 42
Northern Africa 71
Northern Pacific Rim 51
Northwestern United States 16, 19, 77, 133-134, 141-142
Nova Scotia 42, 44

nuclear energy 157-159, 169, 171, 180, 182
nuclear holocaust 135
nuclear weapons tests 182
nuclear winter 158
nuclearization 134
Oakland 73
Ocean convection patterns 43
ocean currents 19, 43-47, 60-64, 71, 74
ocean memory 42-43, 60
ocean pollution 25
ocean temperature 19, 48
ocean memory 42
OECD 181
Ogallala Aquifer 66-67
Ohio Basin 28
oil and coal 9
oil tanker accidents 43
Oklahoma 66
Omaha 72
onset of puberty 107
oppression of dissent 134
orchids 88
Oregon State University 70
organic matter 15, 30
OSU 70
overconsumption 186
overplowing 109, 114
overpopulation 153
overpumping 114
oxygen 9, 15-18, 26-34, 39, 80-85, 89, 143, 177
ozone 9-10, 16-19, 25, 28, 38, 73-86, 100, 121, 128, 143
Pacific Gas and Electric 171
Pacific Northwest 16, 19, 77, 133-134, 141-142
Pacific Rim 63
Paine 137
Pakistan 182
Palm Beach 54
Papua 119
parasites 98
partial nuclear test ban treaty 76
passive solar design 167
Patrusky 16, 77
peace 116

pendulum phenomenon 37
permaculture 185
Persian Gulf 144
pesticides 134
Philadelphia 55
phosphates 111
photolysis 27
photons 78
photosynthesis 27, 98, 105
photovoltaic cells 120
Photovoltaic prices 170
photovoltaics 120, 169-171
phytomass 88
plankton 16, 25, 30, 32, 38, 81-82, 128
plants that are protected by birds 88
plutonium 158
poachers 138
polar icecaps 45
Polynesia 59
population 17, 93-94, 102-108, 112, 116-121, 126, 134, 137, 140-155, 169, 175-179, 189
potato leafhopper 68
Potomac River 56
poverty 140, 148, 154
power plant rehabilitation 160
prairie grasses 65
precipitation 14, 20, 35, 37, 44-45, 69-70, 89, 97-99
Providence, R.I. 55
Prudhoe Bay 62
Puerto Rico 141
pyramid of wealth 136
radiant energy 8, 78
radiation 4, 7-8, 14-16, 28-33, 38-39, 78-85, 158, 161, 167
radioactivity 43
rainfall 19, 21, 58, 71, 79, 94, 141
rainforests 89-95, 115-119, 128, 133, 138, 141, 155
rate of monetary exchange 174
Reagan 135
reflected heat radiation 7
reforestation 98, 100, 126, 147, 183-185
renewable sources 171

Revelle 9
rice 11, 58, 110
Rocky Mountain Spotted Fever 69
Rondônia 94-95
Rosenfeld and Hafemeister 168
rosy periwinkle 101
Rousseau 129
Rowland 16, 78-79
rubber 92
runaway 33, 39, 60
runaway greenhouse reaction 25
Sagan 32-33
Sahara desert 8, 19-21
salinity 47, 50
saltwater intrusion 57, 72
San Francisco Bay 49, 51, 73
San Francisco earthquake 73
Santa Ana winds 72, 74
Santayana 132
Sargasso Sea 41-42
satellites 4, 10, 64, 77
Saturn 26
Scandinavia 19
Schecter 167
Schneider 23, 45, 56
Schneider and Chen 52
Schneider Index 23
Scotland 42
screw-in fluorescent 160
Scripps Institute of Oceanography 9
sea ice 14, 44-45, 51, 65
sea level rise 14, 30, 45-60, 179
sea turtles 100
sea walls 48
seagoing waterfowl 100
Senegal 152
sensitivity to initial conditions 36
Sichuan program 149
silicon 26, 169
small family farms 178
Smith and Tirpak 46, 68, 73, 100
smog 80
social engineering 136, 185
soil minerals 31, 91
soil moisture 20

soils 18, 20, 30, 61-65, 74, 89-91, 109-112, 128, 140, 176, 182
solar architecture 167
solar cycles 28
solar water heating 171
solar-powered cars 163-164
solar-wind hybrid electric vehicle 163
Solomon 16
South Dakota 66
South Pole 16, 38, 60, 75, 77, 79
Southeast Asia 64
Southeastern United States 97
southern latitudes 5, 100
southern oscillation 63, 64
southern pine 99
Soviet Union 19, 109, 143-144, 149, 182
soybeans 68, 116
Spain 171
species 19
species extinctions 14, 40, 47, 96, 101-102, 134-139
spruce 97, 99
Sri Lanka 59
St. Lawrence River 45
Star Wars 135
status of women 152
stealth bomber 182
storms 4, 36, 47-59, 65-73, 113
Subak 142
Suess 9
sugar maple 97, 99
sulfur 9, 26-27, 120, 174
sulfur dioxide 9
Summer of '88 22-23, 43
sun 4, 7-8, 26-42, 61-63, 72-80, 89, 115, 140, 167, 170, 173, 178
Sunraycer 162
superconductivity 168-169
supercontinent of Pangea 96
supergreenhouse 33
superinsulation 120, 168
superinsulating windows 168
supermagnets 168-169
surface temperature history 61

Suriname 57, 58
Suruí 155
Susquehanna 55
sustainable agriculture 176, 178
sustainable development 156
Taiwan 63
Tanzania 119
Tao of Nature 134
Tasmania 63
telecommunications 165
temperate forests 88-89, 97
Texas 66, 100
Texas City 54, 55
Thailand 181
The Hague Declaration 145
The Philippines 63, 93, 141
The Sahel 71
Third World 57, 91, 93, 134, 139, 148, 175-176, 182
Thoreau 132
tick-borne diseases 69
time delay 17
Tirpak and Smith 46
Titan 26
Titus 53
tobacco 115, 175
Tokelau 59
Tokugawa Ieyasu 106
Tokyo 51
Tonga 59
topsoil loss 177
toxic waste 43, 82, 145
trace gas chemistry 9, 11, 14
tree respiration 30
tree-years 188
trees that are protected by fungi 88
trees that make their own insect repellent 89
trihalomethane 72
Trinidad and Tobago 59
tropical forests 91-92, 141, 185
troposphere 85
Truman 135
tundra 16, 30, 38, 62, 88
Turkey 181
Tuvalu 59

typhoons 48, 70
U.N. Environmental Program 144
Uganda 152
Ukraine 70, 91
ultraviolet radiation 38, 78-85, 161
UNEP 144
United Helping Hand 161
United Nations 93, 140-151, 176, 184
United States 119
University of East Anglia 6
up-the-stimulus lifestyles 134
Uranus 26
urban heat island 71
urban mass transit 165
Uru-Eu-Wau-Wau 155
vacationers 164
vanishing paradigm 128
variation in orbit 29
vegetarian diet 116
Venezuela 59
Venus 7-8, 26-27, 34, 39
Vietnam 180
Virginia 100
volcanic 9, 33, 45, 59
volcanoes 33, 91
wan-xi-shoa 149
warped railroad tracks 72
Warsaw Pact 182
Washington D.C. 56, 71-72
water vapor 27
wavelengths 7, 83
weeds 68
Weser Valley 28
West Antarctic Ice Sheet 38, 46, 52
West Germany 19, 118
Western Ecuador 101
Wetherald and Manabe 70
wetlands 17, 47, 56-57
Wheel of Fortune 141
wilderness 116
wildlife 47-49, 55-57, 65, 117, 140
Wilms 49, 72

Wilson 96
wind turbines 171
Windward Islands 42
winged bean 101
Wirth 22
Wiseman and Longstreth 69
Woods Hole Marine Biological Laboratory 112
World Bank 18, 93, 95, 149
World Commission on Environment and Development 101
world energy growth 160
world irrigated area 113
world military spending 179
World Resources Institute 93
Worldwatch Institute 93, 108-113, 119, 149, 150, 162, 170, 173, 179, 181-182.
Wunch 43
Yang-shao 20
Yanomami 155
Younger Dryas 44-45
Zambia 91, 119
Zoró 155